Buffett's 2-Step Stock Market Strategy

Know When to Buy A

Stock,

Become a Millionaire,

Get The Highest Returns

Copyright © 2020 by Danial Badruddin Jiwani

All rights reserved. No part of this publication may be reproduced, distributed, or transmitted in any form or by any means, including photocopying, recording, or other electronic or mechanical methods, without the prior written permission of the publisher, except in the case of brief quotations embodied in critical reviews and certain other noncommercial uses permitted by copyright law.

Dedication Page

This book is dedicated to Barkataly Jiwani, my grandfather.

Table of Contents

AN IMPORTANT INTRODUCTION	5
THE ULTIMATE PRINCIPLE OF INVESTING	11
STEP 1 IS FINDING A WONDERFUL BUSINESS	23
STEP 2 IS DETERMINING A WONDERFUL PRICE	63
MAKING DIVIDEND INVESTING POWERFUL	107
WHEN DIVERSIFICATION HURTS	125
EMOTIONS KILL YOUR RETURNS	141
MARKET CAPITALIZATION AND OPPORTUNITIES	155
TAKE THESE STEPS BEFORE INVESTING	165
WHEN TO SELL A STOCK	175
CONCLUSION	184

An Important Introduction

It would be nice to say that I came up with all of the ideas in this book, but I did not. It is very hard to create a new invention and build new principles. Especially when the fundamentals of investing 100 years ago are the same today, it is very difficult to come up with a principle that has not been discovered.

This book is not focused on reinventing investing or creating a new, unproven strategy. It will not uncover ideas which the world has never seen. It is not going to develop theories about technical analysis and uncover stock chart patterns.

However, this book uses fundamental investing strategies that have been used over the past century. The book takes the logical investment principles, which should be known to investors, and reveals them to you. These are the principles that make sense intuitively and do not require any sort of college degree to understand.

Books are tools for democratizing information to everyone, regardless of whether you have a Ph.D. in finance or even don't have a high school diploma. Whatever education level you have, the principles of investing are possible to learn and do not require a certain degree. I am 17 years old and I can do it. You can do it to. Whether you have lots of money or have very little, the principles of investing don't change. You have the ability to take control of your money and make it work for you.

This book will provide you a 2-step framework to pick and choose stocks so that you will outperform your friends and Wall Street. It will share only 1 strategy. There are many investing strategies that people have invented. However, you only need to know 1 amazing strategy so that you can be successful in the stock market. So, this book may seem shorter and more concise than other books because it will focus only on 1 strategy. If I teach you many strategies, I would be doing an act of disservice to you. It is much better to have 1 amazing strategy that predictably provides good returns without excessive risk than many mediocre strategies that work sometimes. So, this book will provide you with 1 strategy that legendary investors like Warren Buffett have been using.

I want you to have the following mindset when you are investing: if you are not beating the mutual fund managers and index funds of Wall Street, then you are losing opportunities.[1]

Anyone who is picking stocks should have the goal of outperforming the mutual fund managers and index funds of Wall Street. If you are not doing better than them, then you might as well buy their stocks by investing in their funds. You don't need to waste time picking stocks if you can end up making more money by investing in an index fund.

You Have More Advantages Than Wall Street

I don't want you to be intimated by the fact that I want you to outperform the people who invest in stocks for a living. You have so many advantages over Wall Street. You should have no excuse to not outperform them if you take the time to learn how to pick stocks.

Mutual fund managers on Wall Street are at a severe disadvantage. A mutual fund manager's goal is to have as many people

---

[1] Mutual funds and index funds are where most people have their money invested. They are a collection of stocks people can invest in to maximize diversification. The managers of the funds are the people who run mutual funds. They typically invest in hundreds of stocks, and they get a safe, average return of 10% per year. In this book, we are going to be referring to funds as "the market" or "Wall Street."

as possible invest in their fund. If more people are invested, they can get more profit because they are managing more money. In order to get more customers, they sell a sense of safety to their customers (i.e. investors). In order to convince potential customers and everyday people that their money would be safe in their fund, they have to invest in literally hundreds of different stocks. These managers have to invest in so many companies so that they can tell their customers "don't worry, your money is invested in hundreds of different stocks, so the risk is spread across many investments.[2] It is a very low risk fund."

This is what I like to call "diversification to the extreme." By investing in so many companies, they are investing in few stocks which are solid investments, many stocks which are "ok" investments, and some stocks which are losers. By investing mostly in average companies, owning very few solid investments, and holding the losers, the fund managers only get an **average** return on investment.

You have the power to invest **only in** the stocks that will predictably and safely be winners. You don't have to diversify to the extreme because you don't have any customers who want safety. You are your only customer, and all you care about is getting good returns without excessive risk. You have the power to only invest in the companies that will outperform the market, unlike fund managers who have to invest in so many companies. You have the advantage over Wall Street, so you should feel confident that you can beat them at their own game.

---

[2] The idea of spreading risk across many different investments is known as diversification. We are going to be diving into this topic in a later chapter.

Also, these fund managers have an odd dilemma. They know the importance of "buying low and selling high."[3] However, it is very difficult for them to do that. When there is fear in the market (i.e. stock prices are going down), everyday people pull money out of managers' funds. So, when the markets are low and when managers want to be buying stocks, they don't have money because the people stopped investing in their funds and pulled their money out. The funds have no cash when they need it most and have the best opportunities to make money in stocks. Conversely, when the markets are going higher and higher, people are investing in their funds and the managers have plenty of cash. However, they do not have good opportunities to deploy large amounts of cash because there are very few good investment opportunities when the markets are high. This causes them to purchase many stocks at "ok" prices and very few stocks at attractive prices. This makes it hard for Wall Street to get good returns.

You, on the other hand, have control over when to invest your money. You can deploy cash when stock prices are low. You **do not** have customers who are emotional and cannot handle volatility and falling stock prices.[4] You are your only customer, and you only need to make yourself happy. If you can make rational decisions, and if you have the

---

[3] If you are new to investing, you are going to learn very quickly that you make money in the stock market by buying stocks when they are undervalued. In other words, you usually buy stocks when their prices are down and aim to sell them when their prices are higher. That is one of the first things you need to know about long-term investing.

[4] The word "volatile" refers to the fact that a stock's price moves up and down a lot. It only refers to how much a stock price moves. It does not have anything to do with the risk of a stock. When investing for the long term, a more volatile stock is not necessarily more risky than a less volatile stock. However, risky stocks tend to be volatile, but not always.

right strategy, you will become a successful investor and beat Wall Street.

After reading this book, you will have a 2-step strategy to pick winners and avoid losers. You will have control and power over your investments. Most importantly, you will have clarity. You will know exactly when to buy and sell a stock. With the right strategy, you can beat Wall Street at their own game.

The Ultimate Principle of Investing: Summary

As the title suggests this chapter focuses on the fundamental principle of investing: you buy a business for its ability to produce cash and for no other reason. Before we go deeper into this book, it is key that you understand that businesses are like real estate. You buy businesses for the cash that they will produce over their lifetime. We will also look into a metric called "free cash flow" and why it is better than net income or earnings.

This chapter provides insight on the following topics. You need to understand these insights to make informed decisions.

- Why P/E ratios are not the basis for deciding whether or not a stock is at a good valuation
- How Warren Buffett buys stocks
- Why net income or earnings is not what you think it is
- Why you invest in stocks as you would buy a business
- How free cash flow is crucial to investing

The Ultimate Principle of Investing

"It's far better to buy a wonderful company at a fair price than a fair company at a wonderful price" - Warren Buffett

One key principle to understand is that you buy a stock like you would buy a business in real life. Buying stock is literally like buying part of a business, so why wouldn't you buy a stock like you would buy a business? So, if you know how to buy businesses, then you know how to buy stocks.

How do you buy a business? First, you must understand that buying a business does **not** entail purchasing it just because it has a low P/E ratio relative to competitors.[5] Almost every single investing book will tell you that you have to look at the P/E ratio of a stock or business and invest in it when the ratio is low. However, this is not the best way to determine if a stock or business is undervalued. In fact, Warren Buffett has said that "the ratio of price to earnings… [has] nothing to do with valuation."[6]

---

[5] The P/E ratio or price to earnings ratio is frankly the most common metric used by long term investors to determine whether or not to buy a stock. It is calculated by dividing a stock's price by its earnings per share. It can also be calculated by dividing a stock's market capitalization by net income. It shows how much you are paying for a stock relative to its net income. In general, low P/E ratios indicate that a stock is more undervalued, while high P/E ratios indicate that a stock is more overvalued. Although this is one of the most popular financial metrics to use, I will explain why I don't like it and prefer to use something else.

[6] If you are new to investing, here are some things to know about valuation. The word "valuation" is also synonymous with market capitalization. Valuation

In order to understand the fundamentals of how people buy businesses and how you should invest in stocks, I want you to imagine the following scenario because it will completely change the way you think about stock market investing. Imagine that you have $1,000,000 which you can spend on whatever you want, and your friend asks you if you would like to purchase their car dealership. You are really interested in cars and think that the business is really strong. There are few competitors in the area, and your friend's dealership has the best service and the highest-quality cars in the area. When analyzing the firm's financials, you notice that their debt is manageable.

Their expected profit is not the best in the short term. Next year they are expected to break even. However, every year after that, the business is expected to make about $500,000 per year in free cash flow.[7] Given that you think that the data represents realistic expectations, you decide to invest in the company because you are

---

and market capitalization refer to the cost to buy an **entire** business. When you are investing in stocks, you are investing in part of a business. You can buy parts of a business through investing in individual shares of stock in the stock market. Each stock has a certain number of shares on the market, and the value of all the shares is equal to a stock's valuation. So the valuation of a stock is equal to the current stock price times the number of shares of stock outstanding. On a day-to-day basis, the number of shares in the market is the same. So, as more people buy a stock, its stock price and valuation go up. If a person says that stock is trading at a good valuation, it means that they think it is undervalued. You always want to buy businesses when they are undervalued. If an entire business is undervalued, it means that its stock is undervalued. In other words, if the entire business is undervalued, then part of the business is undervalued, and it makes sense to purchase its stock. That is important to understand.

[7] Free cash flow is similar to net income, but it is not exactly the same. Free cash flow tells you how much cash a business generated over a certain time period. It is a very important metric.

expected to make your money back in three years; the business is expected to generate plenty of cash relative to the price to acquire the business (i.e. purchasing the business would lead to a good return on investment).

It is important to notice *how* it was determined that the car dealership was a good investment. In the scenario, the *free future cash flows* of the business were looked at to determine whether the business was a good investment. It seems pretty simple. You looked at whether the business would generate lots of cash relative to the price to acquire the business (i.e. you looked at whether the cash that the business would generate would lead to a good return on investment).

However, most people ignore this fundamental investing principle of looking at a firm's *future free cash flows* when they start looking at the stock exchange. People don't make investing decisions based on a stock's future free cash flows, and they simply look at whether a business will be trading at a higher price in the next few years (i.e. most investors just look at whether they would be able to sell the business/stock at a higher price within a few years). But this is simply not why anyone buys a business in real life. You don't buy a business to resell it after a few years. You buy a business because it produces cash.

Never forget this principle of long-term value investing: you buy a business because it is cheap relative to its future cash flows, and you do not buy a business because you think that you can sell it higher at a higher price to someone else.

I was recently talking to one of my friends. He is very new to the stock market, and the most he knows about investing is what he learned from watching a few YouTube videos. He asked me if I could take a look at his portfolio. I took a look at his portfolio, and I honestly did not recognize any of the companies that were in his portfolio. I frankly believed that he probably did not recognize the stocks that he owned, so I asked him the most important investing question: "why did you buy these stocks?"

"Their prices were low, and I bought them because their price will go up," he said.

His answer was very logical, but also very wrong. I do not blame him for buying a stock because he thinks that the price will go up after it went down, but that is not a good reason to buy a stock. You are probably wondering the following. Why is this a mistake? Why is it bad to buy a stock just because you think the price will go up after it went down?

Well, listen to what Warren Buffett said in an interview about buying a stock because you think that it will be trading at a higher price at a future date:

"Just looking at the price of something, you're not investing. I mean if you buy something, Bitcoin for example, or some cryptocurrency, you're not looking to the asset itself to produce anything. If you buy an apartment house, you're looking at how the apartment house does. If you buy a farm, you look at how the farm does. If you buy a whole business, you're looking at how the business does. If you buy a part of a business, why shouldn't you look at how the business is going to do?"

Warren Buffett is saying that a change in a stock's price is not indicative of a good investment. An investor has to look at the cash that the business produces. For example, as Warren Buffett says, when an investor is investing in real estate, the investor looks at whether the property is going to generate enough cash to make it a good investment. The investor would estimate the amount of rent he or she can collect from the tenants. The investor would then estimate and subtract annual costs related to owning the property. Taking the difference between the rent and expenses would be the net cash that the rental property will generate (this is like free cash flow for a business). The investor would look at their estimation of the property's future free cash flows and compare it with the cost to acquire the property. This would allow an investor to determine if the cost to purchase the property is low enough to get a good return on investment. It is the same way you determined that the car dealership, a business, was a good investment.

An investor would not compare the cost to acquire the property and compare it to the price that he thinks that he can sell it for. When an investor tries to buy a stock or any investment only to sell it to someone at a higher price, Warren Buffett says that he is playing a game of "who beats who" and that the investor who simply looks at price is "not investing."

In essence, investing in a business is like investing in real estate. When purchasing a rental property, you would naturally look at the cost to obtain the asset and compare it to the amount of cash that it will generate. If you were privately acquiring an entire business, you would compare the cash that the business will generate in the future and compare it to the cost to purchase the business. If you compare the ability to generate cash with the cost of an asset when investing in real estate or buying a whole business, why wouldn't you follow the same process when investing in part of the business (i.e. in stocks)?

A T.V. show that I like to watch is called Shark Tank. It airs on CNBC, and it is a show where business owners pitch their business to a group of investors. The business owners seek to gain an investment, and the investors seek to gain some equity or ownership in the business. Some entrepreneurs own a business that has not made a profit yet. On the show, whenever I see an entrepreneur trying to pitch a business that has not generated cash, the outcome is almost always the same; the entrepreneur does not get a deal nor investment. Why? The business has not made money, (and won't likely make money in the future) so the business won't provide a return on investment for its owners.[8] It would be like owning a rental property where you have no tenants but you still have to pay taxes and expenses related to the property. It is not a recipe for successful investing.

When you make decisions solely based on price and not on future free cash flows, you are putting a lot of risk on yourself. I received an email from a reader after I published one of my first books, *The Beginner Investor*. For privacy's sake, we'll call him Ronald. Ronald had a problem. He needed to find a way to effectively deploy his money in the stock market, but he also wanted to outperform the market so that he could retire with lots of money. On his own, he studied investment strategies, and someone recommended a strategy that involved buying stocks based on price movements. For example, he would look at stock charts and guess whether the price will be higher or lower in the short term. It was probably one of the worst mistakes of his

---

[8] Sometimes entrepreneurs do get a deal with an investor, despite the fact that the business has not made a profit yet. The reason for the investor to make the deal is as simple as the investor sees potential in the business to become profitable fairly quickly. However, it is unlikely that an entrepreneur would get an investment without having made money because it is hard to determine **how** profitable a business will be in the future if the business is not currently generating cash; an investor would not have any data to base his or her estimate of future free cash flows off of. It would be like buying a property without knowing the amount of rent you can collect from it. Therefore, the idea doesn't seem promising.

life. He initially invested $200,000 in the stock market. However, just after a few years, his stocks were only worth $30,000. He was down 85% over a couple of years[9].

He would have never sustained such heavy losses if he adhered to the tenants of this book. By understanding the principles in this book, and by focusing on future cash flow, you will never be in such a bad position. In fact, this book will allow you to outperform the market and make more money than most investors. Nonetheless, at the time that Ronald contacted me, he had already lost 85% of his money, and he just wanted help properly deploying his capital. I gave him some of my thoughts on what I would do in his situation, and I told him about how index funds are a much safer investment than individual stocks. Now, his capital is properly deployed, and he is currently retired and has financial security.

Why Net Income and P/E Ratios are Misleading

The P/E ratio is one of the most commonly used metrics to evaluate a stock. Whenever you get a quote for a stock, it is one of the very first metrics you will see.[10] Many investors base entire investment decisions

---

[9] I rarely see investments go this terribly. This is a more extreme situation, but it is still a real example of someone losing lots of money in the stock market.
[10] The P/E ratio or price to earnings ratio is simply the stock's price per share divided by their earnings per share or the stocks market capitalization (valuation) divided by their net earnings. Low P/E ratios tend to indicate a stock is undervalued, while high P/E ratios tend to indicate that a stock is overvalued.

using the P/E ratio. However, this is not the best metric to base your investing decisions off of.

So, why should you not make entire investing decisions based on the P/E ratio? The main reason is that the "E" in P/E ratio does not represent the company's ability to generate cash. This can sound confusing at first. Why would the earnings of a company not represent the amount of cash that they generated?

In order to understand the weaknesses with the "E" in P/E ratio, you must first understand some basics of Generally Accepted Accounting Principles (GAAP) accounting. The Generally Accepted Accounting Principles are accounting rules which standardize accounting. The rules ensure that the profits, assets, and liabilities at a company were calculated the same way at another company. These rules minimize the chances that a company misrepresents its profits, assets, and liabilities, which makes them very important in creating trust between investors and companies. Without the Generally Accepted Accounting Principles, investors would lose trust in the stock market as they would be unsure of whether a company is calculating their metrics the way other companies are calculating them.

The generally accepted accounting rules for calculating net income and earnings per share do not reflect a stock's ability to generate cash. Net income is calculated by including *non-cash charges*. Non-cash charges are things that change a company's net income, but they result in no change in the amount of cash that a company generates.

For example, one type of non-cash charge is known as depreciation. In **accounting**, depreciation is **not** a decrease in an asset's value, and it is **not** a decrease in a stock price's value. Rather it is an expense that occurs when a company purchases an asset that they will use for longer than one year. The expense is recorded over the useful life of the asset due to the revenue matching principle.[11] The revenue matching principle essentially says that a business has to record expenses associated with an asset in the same time period that the asset is used to generate revenue. For example, if a company buys a truck, and they plan to use the truck for 5 years, the company would need to record the expense of purchasing the truck over the span of 5 years. If the company paid $100,000 to purchase the truck, they would record $20,000 in depreciation expense per year over 5 years, even though no cash will be actually leaving the business in those future years.[12]

Why does this matter for an investor? Well, let's assume that the truck generated $100,000 in revenue in 2020, was purchased in 2020, had a useful life of 5 years, and that the only expense associated with the truck was the $100,000 to actually purchase the truck. According to GAAP accounting, the truck would have made $80,000 in net income in 2020 (It generated $100,000 in revenue, while it only had $20,000 in depreciation expense because the $100,000 cost to acquire the truck would be spread over its useful life of 5 years). However, the truck would have really not generated any cash in 2020 because the truck did $100,000 in revenue, while the company paid $100,000 to acquire the truck. In reality, $0 entered the business in 2020, even

---

[11] The useful life of an asset is just how long an asset is expected to be used by a business, and depreciation is simply the process of spreading the cost of an asset over its useful life. Although a business may spend $X on an item today, its expense could be recorded for many years into the future, even though no cash will leave the business in future years. When depreciation expense is used to calculate net income, net income does not represent the amount of money that a company generated in a certain year.

[12] This example assumes a $0 salvage value and uses the straight-line depreciation method in order to make the example simple for the purposes of this book.

though their net income would be $80,000. Therefore, net income does not accurately represent the amount of cash that a business actually makes.

Since many large businesses on the stock exchange have billions (or millions for smaller businesses) of dollars in non-cash charges, their net income can vary significantly from the amount of cash they are actually generating. In fact, I have seen situations where a business has positive net income, but they are not actually generating cash.

Since the "E" or net income/net earnings in "P/E ratio" does not represent the amount of cash that the company generates, the P/E ratio does not represent the cost to purchase a company relative to the cash that it generates. The P/E does not accurately allow investors to determine whether a stock is a good investment because it does not look at a company's actual ability to generate cash. Since investing is always about looking at an asset's ability to generate cash, the P/E ratio should not be the focus of investment decisions.

However, I do not want to entirely disregard P/E ratios. Lower P/E ratios indicate that a stock is undervalued. That is 100% true. But, where the P/E ratio comes short is telling investors when to buy a stock. When I watch people give reasons to buy a stock, they say that it is at a "low P/E ratio." But how low does a P/E ratio need to go in order to make a stock a good buy? What is "too high" of a P/E ratio? When you rely on P/E ratios for investing, there is more guesswork than certainty, especially since the P/E ratio does not compare the price of a stock, relative to its ability to generate cash. So, I would rather make investing decisions by looking at the cash that a business will generate and compare it to the price of the business. Then, the answer is much more clear than simply looking at the P/E ratio of a stock.

The Ultimate Principle of Investing: Contract

Instead of simply telling you how to make money in stocks, I want to help you make money in stocks. So, at the end of each chapter, there is going to be a contract. It will outlay the key points of each chapter. I want you to sign your name on the line that is below. The contract will help you be more committed to the tenants of investing[13].

I _____ (insert name) agree to the following principles of investing:

- I will invest in businesses for its free cash flow
- I will invest in stocks as I would invest in a whole business or a piece of real estate
- I will not make entire investing decisions based off of the P/E ratio

_____

[13] In Robert Cialdini's book called *Influence*, he describes a marketing principle called "consistency." He essentially describes how people are more likely to follow through with something if they agree to it earlier. So, signing this contract will help you stay committed to the principles of investing.

Step 1 is Finding a Wonderful Business: Summary

In order to become a successful investor, you need to just follow the 2-step investing framework. The first step is to find a wonderful business. Being able to identify whether a business is wonderful or not is a key step to investing. It will ensure that your business will obliterate the competition. A wonderful business has 4 characteristics: it has a moat, operates in a slowly changing industry, has predictable free cash flows, and has strong financials.

Topics in this chapter include:

- Why you should not invest in the next big thing to be successful at stocks
- How predictability is your best friend when investing
- The 3 types of moats that make businesses wonderful
- A moat's most important quality
- How to use the income statement, balance sheet, and cash flow statement to your advantage
- A deep look into the meaning of free cash flow

Step 1 is Finding a Wonderful Business

"The key to investing is not assessing how much an industry is going to affect society, or how much it will grow, but rather determining the competitive advantage of any given company and, above all, the durability of that advantage." - Warren Buffett

This chapter is one of the most important sections of the book, so you should not skip over it. It will focus on the characteristics of a wonderful business so that you know which businesses are potentially good to invest in.

Being able to identify wonderful businesses is probably the most important step of investing. I would even argue that it is more important than the second step of investing: investing at a wonderful price.

We are going to take a look at the 4 characteristics of a wonderful business which include having a moat, operating in a slowly changing industry, having predictable free cash flows, and having strong financials.

If you find a business that has all 4 of these characteristics, you know that you have found a wonderful business. Once you have found a wonderful business, it is then time to determine the perfect price to invest in the business. But, for now, let's focus on identifying the characteristics of a wonderful business.

Characteristic #1 of a Wonderful Business: It has a Moat

I want you to imagine that you just purchased a new house that has nothing in it except for a precious pile of diamonds. Your new house does not have any security system, doors, nor cameras. You want to protect the diamonds because they are very valuable. However, word gets out about the diamonds in your house, and a robber is coming to steal them. As of right now, the robber would be able to walk right into your house and steal the diamonds. However, you don't want to let that happen, so you start to build up your defenses in order to compete against the robber. In order to stop the robber, you first purchase a door for your house. Although it would make it hard for the robber to get in, you know that he would still be able to break in by picking the lock. Then, you decide to set up a security system, which makes the job risky for the robber. And lastly, you purchase security cameras so that potential robbers think twice about entering your house.

Every time, you added a security measure, you made it harder for the robber to steal your diamonds. The diamonds are like the profits for a business, and the house is like the business itself. Every firm's goal is to protect their profits and grow them; most homeowners' goal is to protect their wealth and grow it. Like a homeowner who can add security measures to protect possessions, a business can add security measures called *moats* to protect its profits.

Every business needs a strong moat because it makes it easier to compete against other businesses that want to steal the firm's profits. If a firm does not have any security measures to fight potential competitors and robbers, it will be much more difficult to continually persist.

An investor needs to be able to identify if a firm has any type of moat because it is ideal to invest in companies that have a solid moat. In the marketplace, three types of moats are commonly found: i) the firm has a strong brand ii) the firm is the designated company for its services iii) the company has competitive intellectual property. If a company does not have a moat, it will be harder for them to consistently grow their profits because their competitors will continually steal market share. When a firm starts producing less and less cash, its stock price usually dips. If the trend persists, the stock may not recover, causing investors (i.e. you) to not outperform the market. Therefore, you should look for companies that have at least one, strong moat.

Moat #1: The Firm Has a Strong Brand

The first type of moat is a brand. Brands are images, ideas, and feelings that customers associate with a business. For example, Amazon customers associate the firm's brand with fast shipping times at affordable prices. Apple's brand promises to deliver sleek, simple-to-understand products to its customers.
  But the most important part of a brand is the trust that is developed with customers. The ultimate goal of branding is to develop trust with customers by always delivering on your promise. Starbucks customers can always trust the company to provide high-quality coffee. If Starbucks comes out with a new product, customers will automatically assume that it is a high-quality cup of coffee. However, if a different coffee shop came out with a new product, their own customers might be skeptical of it. It would be harder for the coffee shop to have their customers try new products because they haven't built trust yet. When you notice that a brand has developed trust with customers, you should be certain that it has a very strong brand and moat.

Building brands and trust provide many competitive advantages to businesses. For example, strong brands can lead to easy sales for the long term.

First off, the best marketers know that the goal of marketing to a new customer is not to generate a sale. New customers are usually not ready to purchase from a random business that a customer does not trust yet. It is just a very ineffective marketing strategy (especially for small and medium businesses) because it wastes a lot of money. Even if a sale is closed, the chances of that are very low. It would be like if someone randomly came up to you and asked if you would like to purchase their new online course. You would likely say "no" or just ignore them.

Instead, the goal of initial marketing efforts is to develop a relationship with a potential customer (i.e. develop some trust and a brand). On social media, you probably have noticed that many ads offer free products or services, whether it be a free pdf, webinar, or consultation. Marketers offer these free products and services in order to develop a relationship with you.

When a company has developed a relationship with a potential customer, it has a brand. A business that has developed a relationship with a large audience can easily sell a product to customers again and again. If your friend said that he came out with a second online course, you would most likely buy it because you have a relationship with your friend. For a company that has a strong brand, it is like the firm's customers are the firm's friends.

A company that has an excellent relationship with its customers is Amazon. They have successfully developed a relationship with customers by creating a sense of trust with customers. They promise fast shipping times at affordable prices for customers. They always (at least from my experience) deliver on their promise. Now I, along with millions of other customers, want to go to Amazon for online purchases.

If I want to purchase a new shirt, I do not go to JCPenney's website first because I have not developed a relationship and sense of trust with them. Instead, I would purchase from Amazon because they have developed trust with me. This sense of trust has made me become a recurring customer for them. When you notice that a company has developed trust and created a strong brand with customers, then the firm will have recurring customers. This means that customers will not be purchasing products from the competition, giving the business a strong moat.

Besides creating recurring customers, strong brands make the competition inferior. For example, Apple and Android offer nearly identical products at nearly the same prices; they both offer phones that have the same basic features and apps, yet over 80% of teens have an Apple iPhone.[14] Most teens choose to purchase an iPhone over an android because of Apple's branding. When someone purchases a new iPhone, they are also purchasing the ability to be "cooler" in society. They are purchasing status. However, with an Android, teens are not getting any benefits from their branding. In fact, teens who use Androids are seen as having a phone that is not in style. Since the iPhone makes people feel modern and trendy, they are able to capture a much larger market share among the youth, allowing them to make more money and prevent competitors from successfully competing.

If a business has a brand, it is very hard for competitors to compete. Brands protect profits and act as a moat. If a company has a strong brand that attracts customers, then it has a very strong moat, making it potentially a wonderful company if it has the other

---

[14] The statistic was provided by an article on Business Insider.

characteristics of a wonderful business. Before we look at them, let's take a look at the other types of moats.

Moat #2: The Firm is a Designated Company for its Service

Some companies just have a natural moat around their business model and do not need to develop a strong brand in order to compete.

What kind of companies are we talking about? These companies are the designated company to provide a service. The key is that they are the only or one of the only companies that provide a certain service in a certain area. For example, if there is only one company in the area that can provide an essential service for a customer, then that business has a natural moat because there is no competition in the area. Customers have to go to that one business if they want that service.

For example, a company called Waste Management picks up trash from people's driveway every week. They are one of the only companies in their area that provide this service. Customers have to use their service if they want their trash picked up. In effect, this company and other firms in similar industries are able to have a stable source of revenue. Every year, they are going to have paying customers who are going to want their trash picked up because there are no competitors who can take away their profits.

Many investors would initially be skeptical of these companies. "What is stopping the competition from coming in? If they were in an amazing industry, wouldn't there be more competition?" These reasons make people skeptical to invest in these kinds of industries. But think about the same skepticisms in another way. These firms have such a strong moat that the competition cannot even enter these kinds of industries. If someone wanted to compete against Waste Management, they would need to spend millions of dollars on trucks, infrastructure, landfills, and other large, one-time costs. After making these large investments, a competitor would hope to get a good return on investment. However, the competition would not even generate much cash because it would be too difficult to acquire customers.

If a new company came up to your door and asked if you would be willing to stop giving business to Waste Management, a company that you have been dealing with for years, and switch to their new company, you would say "no." You don't want to turn your back on the company that has been providing a service to you for years. You don't want to go through the hassle of canceling your current service and working with a new business. You would not want to take risks on a new business, especially if Waste Management has been providing their service consistently and fulfilling your expectations.

In business, there are two factors that cause you to choose one product over another: its perceived quality and price. In these industries where a single firm is the designated service provider, a new competitor cannot increase the quality of their service nor charge a lower price in order to compete. First of all, they can't create a higher quality service.[15] The results of the service are the same wherever you go, and everyone in the industry provides it at pretty much the same quality. You don't ever say that "I wish I worked with X waste management company because I love the way that they pick up the trash." Since the service is standardized in these industries, a new business can't offer the service in a better way that would greatly benefit customers. Therefore, businesses can't incentivize customers to switch providers.

---

[15] You are going to find that in most industries where a firm is a designated product or service provider, the business is typically mundane and provides a standardized service.

The only other way that a new competitor can realistically acquire new customers is by offering their service at a significantly cheaper price. But this is also unlikely to happen. Competitors already know that they cannot offer a cheaper price because they have to make an adequate profit in order to make this business venture worth their time. Essentially, **when you notice that a company is one of the only providers of an essential, standardized service in an area, you know that they have a moat** because it is essentially impossible for a competitor to attack.

Some businesses and industries that have this type of moat include:

- The waste management industry
- Certain regulated water and gas companies
- Some defense contractors
- Certain telecommunications businesses
- FedEx and UPS

Typically, businesses that have this kind of moat are not fast-growing or exciting companies. Most of these companies tend to be very stable, slow-growth companies. None of these industries are going to gain or lose many customers rapidly.

For many investors, this is an automatic turnoff. Some investors completely avoid slow-growth companies because they "won't be the next big thing" or "won't see the company's profits grow exponentially." However, the lack of growth can be the greatest weapon for an investor. The stability allows us, investors, to easily predict the amount of cash that the business will generate in future years. It is very easy to predict the amount of cash that the business will generate in the future because the competition, industry, and number of customers barely changes each year. Being able to predict the amount of cash that a business will generate is key. If you know how much cash a business will generate, then you can determine if the price of the stock is overvalued or undervalued relative to its ability to generate cash. It is like having a tenant agree to pay a fixed amount of cash every month. This is a good thing. If you can't determine the amount of money that a business will make, then it is a risky investment because you can't determine whether you can get a return on investment. Therefore, these types of companies can provide some of the most predictable returns in the stock market. If you find one of these companies, it has the potential to be a good investment because they can predictably be good investments.

Moat #3: The Firm has Competitive Intellectual Property

The third type of moat is any form of intellectual property. Intellectual property is broadly defined as any unique piece of work. These can include software, designs, music, and products. When a product, process, or creation has a unique characteristic, it can be protected by the government, giving the creator of the intellectual property the sole right to sell and distribute it. Common types of protections include patents, trade secrets and copyrights.[16]

These are some of the strongest moats for businesses. For businesses offering a product, this is very important. It prevents a firm's product from turning into a commodity because it gives a certain

---

[16]Trade secrets are not protected by the government, but they are a moat because having a trade secret prevents someone from creating a similar product. No one knows the recipe for Coca Cola, a trade secret, giving the company a strong advantage.

business entity the only right to sell a product. If Apple did not have intellectual property and everyone in the world could make and sell iPhones, it would just turn into a commodity. Then, Apple would have to compete with other businesses selling their own product. The competition would inevitably cause the price of the iPhone to fall. Apple would generate much less cash due to the competition and lower iPhone prices.

Another company that has a strong moat due to intellectual property is Activision Blizzard, a company that publishes video games. All of their video games are protected by copyright law, so no other business can produce a copy of their games without Activision Blizzard's permission. Since they own large titles like Call of Duty and Overwatch, it is significantly harder for competitors to take away their customers; no firm can create an identical game and deliver the same experience. If competitors try to create a knock off version, it would be hard for it to get as much traction as the original because customers would just view it as a copycat. This allows Activision Blizzard to have their own group of loyal customers, which the competition would have a hard time taking away. It makes it difficult for the competition to thrive.

Thus, intellectual property defends businesses against the competition, making firms with intellectual property possibly a good investment. However, not all intellectual property or moats are created equal, so it is important to assess the quality of a moat.

The Most Important Quality of a Moat

What is more important than having a moat is identifying the strength of the moat. You only want to invest in businesses that have strong moats. You want to surround your treasure with a high level of security.

In order to assess the strength of the moat, ask yourself the following question before buying a stock: does this company have a stronger moat than the competition?

It is important that you ask yourself that question. This question is key for you to determine the strength of a moat.

There are many companies that have a moat, but many of them don't have a strong moat. If you are thinking about investing in a stock, but there is another company that can realistically outcompete them, then the stock does not have a strong enough moat (i.e. a company can have intellectual property, but the competition's intellectual property could be better and more competitive in the marketplace). You only want to invest in businesses that have a strong moat or else the competition will be brutal for the company. A moat is strong enough only if it prevents the competition from being a threatening competitor.

Let's say that you are thinking about investing in a generic donut store. In order to assess the strength of their moat, you would think about the company's competitors and whether they can successfully outcompete them. After researching the company's competitors and doing your analysis, you may conclude that the donut store would not be able to compete as well as Dunkin Donuts because they have better branding and donuts than the generic donut store. By thinking about the generic company's competitors, you would have found a company that is stronger and has a better brand and a stronger moat: Dunkin Donuts. Since the generic donut store has a weaker moat than Dunkin Donuts, you would not want to invest in it because you only want to invest in strong moats; weak brands and intellectual properties are pointless because they do not bring customers and protect against the competition. By identifying that the generic donut store had a weak moat, you would have avoided investing in a company that would have gotten obliterated by the competition. Also, by analyzing the competition, you would have potentially found a company with a better moat that could be a better investment opportunity. The lesson here is to always analyze the competition in order to see whether a company that you are investing in has a stronger moat.

There is absolutely no point investing in a company with a weak moat. If a company has patented a product that is not good, it won't sell. If a company has a secret recipe or trade secret for food that does not taste better than the competition, it would be hard to sell. A brand has no effectiveness if no one has ever heard of it and if everyone loves the competition's brand. So, it is always very important to look at whether the core product (that is protected by intellectual property) is high-quality and whether the brand is competitive enough to make customers want to purchase a product from them.[17] Intellectual property and brands are only worth investing in if they are better than the competition. So, you always have to analyze the quality of a moat, not just whether a company has a moat. Most businesses have a moat, but most businesses do not have a moat that is stronger than the competition. You want to own strong companies, so invest in the strongest moats.

---

[17] The only exception to the guidelines that a company needs a strong product/brand is when they provide a designated service and have little competition. If there is only one company that can provide a certain service in a certain area, customers have to go to them if they want that service. In other words, there is no competition. The lack of competition is their moat, and they don't need a better brand, product, or service to attract customers.

Characteristic #2 of a Wonderful Business: Its industry is not Rapidly Changing

Wonderful investment opportunities are not found in rapidly changing industries. These industries can be very risky for you. Here are some of the characteristics of a rapidly changing industry:

- A company's product or service can become outdated within a few years, so they have to innovate.
- **If they don't successfully innovate, the competition will take over their customers**.
- Customers only care about having the newest and highest-quality products.
- **Brands don't drive customers to purchase a product** in these industries. Only product quality drives customers to purchase a product.
- **Businesses that have the most modern product will always outcompete the competition.**
- These are typically known as the "next big thing."

Here is my definition of a rapidly changing industry: a rapidly changing industry is an industry where the future outcomes of businesses are unpredictable. **We don't know who will be the best competitor in the long run.**

With these businesses, technology gets outdated after a few years because competitors develop a better product, and the innovation completely destroys a firm's moat. Although a business may have a patent (a type of protection for their intellectual property) on their technology, a competitor can create a better product within a few years, patent that new technology, and steal most of the market share.

When there is rapid change in an industry, I like to call it a "technology war." Companies are consistently innovating in order to keep their customers. You do not want to be invested with a company that is in a technology war. There is just too much risk. One year the company will thrive and attract customers because they have the best technology, and another year they could be struggling because the competition has better technology. You want to invest in businesses that are predictably the strongest competitor. Predictable businesses generate predictable amounts of cash. When a company generates a predictable amount of cash, you can determine whether it will generate enough cash in the future to get a good return on investment; in other words, you know how much rent your tenants will be paying.

You can oftentimes find a rapid pace of innovation within **certain** industries in the technology sector. For example, *some* generic companies which sell and create software are in rapidly changing industries. The problem with these *generic* software companies is that their products might not be relevant in 5-10 years from today because the competition will make a better product. Then, the old business is going to lose customers and significant cash-generating ability. There is simply no guarantee that a business in a rapidly changing industry will have the best product in the long term.[18] Thus, companies in rapidly-

---

[18] It is important to understand that not all technology companies are in a rapidly changing industry. You know that an industry is rapidly changing if you can't easily say that X business will be one of the most competitive businesses in their industry many years down the line. On the other hand, most people can agree that a technology company like Apple will have the best smartphones for many decades. In other words, Apple is not in a rapidly changing industry.

changing industries don't have long-term moats, making them a risky investment.

Warren Buffett has said that, in a business, he is looking for a "long-lasting moat." The keyword is "long-lasting." You want your companies to have a moat that would ideally last for many years. If a company's moat only lasts for a few years, then the company would be vulnerable in the long-term. If you can't assure yourself that a company will have a moat in the long term, then it is very risky to invest in it because the company could be very vulnerable to the competition in 5, 10, or 30 years from now. Since moats in rapidly changing industries do not last for a long time, investing in such industries is risky.

However, I don't want you to assume that innovation is bad. All businesses need to innovate and come out with new products. It is safe to be invested in companies that innovate at a steady rate. For example, major restaurant chains occasionally come out with products to attract new customers. This type of innovation is normal and beneficial for all businesses and investors because it leads to more customers and revenue.

You only have to be worried about an innovative business if the business will fail if it does not consistently make better innovations than competitors. For example, a chipmaker for computers will fail if it cannot ensure that its chips are faster than its competitors' chips. This would make investing in a chipmaker risky because the firm would lose a lot of customers if it fails to innovate one year; you are betting that they can innovate faster than the competition, which you can't predict with certainty.

On the other hand, a business like Apple can still sell its phones if Samsung innovates and has a better camera because Apple has a very strong brand. The company is not dependent on developing and innovating products quickly because their brand is strong. Although they do continually come out with new products, people would still buy them even if they were slightly inferior to Samsung's products because Apple has a brand. Similarly, Starbucks does not need to continually release new products in order to stay ahead of the competition. They can continue to offer their coffee for decades, occasionally come out with a new product, and be fine for decades with little innovation. They are not dependent on innovating quickly in order to have customers. The competition would not be able to take over if Starbucks fails to innovate. The lesson here is that it is safe to invest in companies that are not dependent on innovation.

It is important to differentiate between a rapidly changing and fast-growing industry because many people think that they are the same concept. They are not. An industry (or company) can be growing at a fast rate with little innovation. For example, Apple is not in a rapidly changing industry. It is very unlikely that Apple would lose many customers if a competitor came out with a better phone. The future for the industry is very predictable, unlike rapidly changing industries where the future is very unpredictable. Despite the company being very predictable, it also has good growth. For example, Apple is expected to grow their earnings at about 10% per year over the next 5 years, according to Yahoo Finance. Therefore, a rapidly changing industry is an industry where the future outcomes of businesses are unpredictable; we can't say who will keep up with the rapidly changing technology and become the best competitor. A slow-changing industry is very predictable, and it can potentially have good growth ahead of it.

Before investing in any business, you should always ask yourself the following question: will this business be dominant over the competition for many years?

Asking yourself the question will ensure that the company you are analyzing has a "long-lasting" moat. If you believe that a business can operate for many decades and is very unlikely to lose to the competition, it is likely not in a rapidly changing industry. Although we can never be certain that an industry will remain roughly the same over the long term, it is a good psychological exercise to go through because it allows you to narrow down your investments to the ones that are in predictable industries and have strong, "long-lasting" moats.

Characteristic #3 of a Wonderful Business: It's Financially Healthy

You always have to make sure that a company is financially healthy.[19] Healthy finances prevent disastrous investments.

For those who are new to investing and business, there are 3 main financial statements: the income statement, balance sheet, and the cash flow statement. They give information on a firm's financial performance, and they can be found in a firm's annual report.

Why You Must Analyze Financial Statements

At the end of the day, the only thing that matters in business is the numbers. How much profit did a company make? How did it compare to

---

[19] "Financially healthy" is a very vague term. So, we will take a look at what you need to look for in order to ensure that your stocks are financially healthy. But in general, being "financially healthy" just means that the company's income statement, balance statement, and cash flow statement tell us that the business is strong.

last year? How much debt did they pay off or take on? How much cash will they generate?

If a business is not financially strong, it is not a good business to invest in. A company could have the most amazing product in the world, but if the numbers don't look good, you wouldn't want to invest in that business. Ideally, you want your business to be generating more cash 3 years from now than whatever amount it is generating today. It won't be able to generate more cash each year unless it is financially healthy.

Also, from a fundamental investing perspective, you always want to invest in companies that can easily last for decades. Companies that are not financially healthy are unlikely to last as long because their ability to generate cash will get constrained at some point in time. Whether it takes 5 or 10 years, if a company is not financially healthy, its ability to compete and grow will be severely limited because it won't be able to generate enough cash to fund growth opportunities. Looking for the right factors in financial statements ensures that you are investing in companies that will make you lots of money. Financial statements are very important!

Financial Statement #1: The Income Statement

For those who are new to investing, an income statement simply shows a firm's net income by comparing its revenues against its expenses[20]. On

---

[20]If it is your first time reading a financial statement, it is important to realize that all of the numbers are listed in "thousands" or "millions," depending on who is providing the statement. For example, Yahoo Finance lists numbers in financial statements in "thousands." So, if the income says that a company had

the income statement, you should look for trends in a company's profit, with good growth year over year; on the bottom of the income statement, where it says net income, you want to see the value go up each year because it generally means that the company is making more money each year. It is a simple but important trend.

The next metric to look at is the revenue. Like with profit, it is important to see that the revenue is continually increasing each year because it shows that the business is continually growing. In some cases, a company's profit can be increasing while revenue is falling because they are effectively cutting costs. Although it is important to minimize costs, it is equally important to see rising revenues in a business because rising revenues show that the business is appearing more favorable from consumers' perspectives or that the business has more pricing power.

The next metric to look at is the profit margin. On Yahoo Finance, it can be found underneath the statistics tab. For your understanding, it is calculated by dividing the profit (net income) by the revenue ($\frac{Profit}{Revenue}$). It represents the percent of the revenue that is profit for the business. For example, if a company did $100,000 in revenue with a 30% net profit margin, its net income would be $30,000. A higher margin is better because the business is keeping more of their revenue as profits. When finding a good stock, you should look for a net profit margin of **at least** 5%.

$1 in profit, it means that they really had one thousand dollars in profit; if it says $230,335 in profit, it means that the company did roughly $230,335,000 in profit; when metrics are displayed in thousands, adding 3 zeros to the end of the number will give you its true value. If the metrics are displayed in "millions," adding 6 zeros to the end of each number will give you its true value.

If a profit margin is below 5%, the company does not have a sufficient cushion. Sometimes, businesses will see costs rise due to unexpected factors that are out of their control. For example, the price of commodities used in manufacturing can go up significantly, leading to a firm's products being more expensive to produce. If the company is just barely making a profit, they will be in lots of trouble if they see their costs rise just by a little bit because it could turn them into a business that is not generating cash. A business with a low profit margin is like a piece of glass; they are both fragile. Since you always want to be safe and certain with your investments, you should always look for net profit margins above 5%.

Financial Statement #2: The Balance Sheet

The balance sheet shows two things: assets and liabilities. Assets are simply valuable things that the company *owns.* These include buildings, cash, and inventory. Liabilities are things that the company *owes.*

Notice how I said that liabilities are "things" that the company owes. They are not always debts, where a company borrows money and has to pay it back with interest. Many investors who are not familiar with accounting assume that all liabilities are debts, but that is not the case. A liability can be anything that the company owes. For example, deferred or unearned revenue on the balance sheet is a liability for a company, but it is not a debt. It is recorded when a company is obligated to perform a service or to provide a product in the future. A business could have received money for a product in advance of sending it. Therefore, they would be liable for providing the product. But that is not a debt where they owe money and interest. Just understand that debt is a type of liability and that all liabilities are not debts. It is a very common misconception that should be cleared up.

The most important type of liability to analyze is debt, when a company owes money and interest to the lender. Out of all types of liabilities, debt can be one of the most beneficial and the most dangerous. Debt allows companies to have immediate access to cash which can help the company grow and increase profits. However, debt can be dangerous for companies if they take on too much of it. With too much debt, a company can lose control of their ability to pay it back. The interest and debt can accumulate, ultimately leading to the bankruptcy of the company.

Good companies keep debt to a minimum and maximize assets over the long term. To determine whether a company has too much debt, investors should calculate the debt to asset ratio. It is calculated by dividing the total debt by the total assets ($\frac{total\ debt}{total\ assets}$). When we are doing this calculation, we are comparing *debt* to the company's assets. We are not comparing *liabilities* to the company's assets. In order to calculate a company's total debt, you have to take the sum of their current debt and their long-term debt.[21] The current and long-term debt of a stock can be found on any company's balance sheet. Once you find the current debt and long-term debt of a stock on their balance sheet, take the sum of the two values. This is a stock's total debt. Then, divide that number by the company's total assets, which can also be found on the balance sheet ($\frac{Current\ Debt + Long\ Term\ Debt}{Total\ Assets}$). This will give you a stock's debt to asset ratio, which is very helpful for analyzing a company's debt.

---

[21] Current debt is any debt that the company has to pay within 1 year, while long term debt is debt they will pay back in more than 1 year. These can be found on a balance sheet.

A lower debt to asset ratio is good. It means that the company is at a very low risk of getting swamped by their debt. It tells us that the company is not heavily financed by debt. Ideally, you would want a stock to have a debt to asset ratio under .35. When it is under .35, the company is in a good position to meet all of its obligations. On the other hand, a debt to asset ratio over .6 is too high. A debt to asset ratio over .6 indicates that the company is too reliant on debt and that it can turn into a disastrous situation if something goes wrong.

Comparing a company's total debt to their total assets allows you to determine whether a company is too heavily financed by debt. In addition to ensuring that a company is not too heavily financed by debt, we need to analyze a company's ability to repay their debt. In order to determine whether a company will be able to pay its long-term debts, it is important to compare a company's debt relative to their free cash flow. You should divide a stock's total long term debt by the free cash flow ($\frac{long\ term\ debt}{free\ cash\ flow}$).[22] As with the debt to asset ratio, a lower value is better. It is ideal that this ratio is under 3. You want to avoid situations where the long-term debt to free cash flow ratio is too high and over 3. When the ratio is too high, it is a red flag. It is a warning sign that the company has too much debt. Too much debt does not necessarily mean that the company will go bankrupt, but it does mean that their ability to grow is significantly constrained. The management would be hesitant to take on more debt when they see growth opportunities if they already have too much debt. Also, too much debt means that they would have large interest payments; the business would have less cash available to grow and pay off debts. We always want to make sure that the business is in a position to pay down their debts so that they don't accumulate.

---

[22] A company's free cash flow can be found on the cash flow statement. We will be breaking down what free cash flow represents later in this chapter.

Looking for companies with a long-term debt to free cash flow ratio that is under 3 will allow you to find companies that are unlikely to have issues paying their lenders.

You also need to look at "total stockholders' equity." It is at the bottom of the balance sheet. Total stockholders' equity is calculated by subtracting total liabilities from total assets. It is crucial that an investor looks for this metric to be increasing every year. If it is consistently increasing, it shows that the company's management is running the business well. You should be worried if you see total stockholders' equity decreasing. A siren should go off in your head. If it is decreasing, the company is not creating more assets than liabilities, which shows that the management is not running the company well. It is a sign of a bad investment.

A Key Metric to Know for All Your Stocks

Return on invested capital (ROIC) is a very important metric to look at for all of your stocks. It is calculated by the following formula:

$$\frac{Net\ Income - Dividends\ Paid}{Invested\ Capital} = \frac{Net\ Income - Dividends\ Paid}{Current\ Debt + Long\ Term\ Debt + Total\ Shareholders'\ Equity}.[23]$$

Most stock research websites will have this calculated for you already. But if you would like to calculate it, all of these numbers can be

---

[23] On the internet, there are different formulas for calculating the return on invested capital. The one that is in this book is the easiest formula to use in order to calculate the ROIC.

found on a company's financial statements. Net income is on the income statement, dividends paid is on the cash flow statement, and current debt, long term debt and shareholders' equity are found on the balance sheet. Being able to do the math is not important. You need to understand how to interpret and apply it. The ROIC is like return on investment. If a business has a 5% ROIC, they essentially made $5 for every $100 of capital they had at their disposal. A business with a high ROIC is very desirable. A high ROIC means that the business is generating lots of money when compared to the capital that it raised/invested. It also tells you how well the company is at investing their money. When a business is able to maintain a high level of ROIC, it shows that the management is able to consistently invest their cash and get a good return on investment. You always want your company to be getting a high return on investment.

    Many of the best investors in the world look for this characteristic in their businesses. Whenever I watch an episode of *Shark Tank*, the investors usually ask the entrepreneur, "how much money did you invest in the business" and "how much money have you made so far and paid yourself." When they ask about the money that the entrepreneur invested so far, they are asking about the entrepreneur's invested capital (how much money they had at their disposal). And, when the Sharks ask the entrepreneurs about how they made and paid themselves, they are essentially asking about the numerator (net income - dividends) in the ROIC equation. They ask about the entrepreneur's invested capital and net income in order to determine the firm's ROIC. Knowing the ROIC allows the Sharks to know about how smart the entrepreneur is at spending and making money. Pretty much whenever I see that an entrepreneur has a bad ROIC, the Sharks are unwilling to invest. I remember seeing one episode in particular where an entrepreneur spent $450,000 on a business idea but did not have any sales. The situation could not have been much worse. Who would want to invest in a business that did not make money after investing over $450,000?

Charlie Munger, Warren Buffett's partner, also highly values the ROIC of a business. Here is what he has to say about the ROIC of a business:

> "It's obvious that if a company generates high returns on capital and reinvests at high returns, it will do well."

Essentially, Charlie Munger is saying that, if a company is able to **maintain** a high ROIC and reinvest their cash in the business, (i.e. not pay dividends) it will do well over the long term.

When a business has a high ROIC, it can compound its growth phenomenally. Let's say that a business is able to have a 10% ROIC each year, pay no dividends, reinvests all their profits into the business, and has $100 in invested capital this year. This year it will end up making $10 in net income:

$10\% \; ROIC * \$100 \; Invested \; Capital = \$10 \; Net \; Income - \$0 \; Dividends$

Since they made $10 in net income this year and are reinvesting all their net income, they would have $110 in invested capital next year. Then, they would get a 10% return on the $110, allowing them to have a net income of $11. They would then reinvest the $11, allowing them to have a total invested capital of $121. In the following year, they would get a 10% return on the $121. I think that you are getting the idea now. Having a high ROIC is like compound interest for a company. *Being able to maintain a high ROIC for many years* allows a company to grow exponentially. In essence, you need to invest in companies that have a good ROIC and that will be able to maintain it at good levels. It also makes an investment attractive because it shows that the management is able to continually get a good return on investment as they manage larger sums of money. Who would not want to invest in a company that has a good ROIC and then gets a high ROIC as the invested capital grows?

To get good returns, it is crucial that you only invest in businesses that have a ROIC over 10% and that *have been able to maintain it at above 10%* (unless you are investing in a dividend stock).[24] A ROIC above 10% is a sign that a company is of very high quality. It shows that they are able to manage large sums of money and get a good return on investment. If your company is always getting a good return, then their stock will do very well over the long term (even if you don't buy at the perfect price). You also want to make sure that it has maintained a high ROIC in the past and that it is not falling. This will allow you to be sure that the business will have a high ROIC in the future. It is important that the ROIC can be maintained at high levels in the future or else your company's growth will slow down substantially.

Frankly, a business cannot be wonderful without a high ROIC, unless it pays a dividend. If I could only look at one financial metric of a company, the ROIC would be pretty high up on my list. It is just so important to look at that you would be doing yourself a disservice if you do not.

---

[24] The ROIC tells you the net income *that stays within the business* relative to the invested capital. When a company pays a dividend, most of their net income is paid out as dividends. Since dividend-paying companies usually payout most of their net income as dividends and since ROIC measures the net income that stays within the business, dividend-paying companies naturally have a very low ROIC. So, it is unrealistic to expect dividend stocks to have a high ROIC.

Financial Statement #3: The Cash Flow Statement

The cash flow statement shows how the business is receiving money and spending it. If you are new to investing and accounting, you are probably wondering how the cash flow statement differs from the income statement because they sound very similar. They both show how the company is making and spending money.

The main difference between the income and cash flow statement is that the cash flow statement strictly tracks exactly how cash is being entered and spent in a business, while the income statement includes non-cash charges per GAAP accounting. We took a small peek at this during an earlier chapter. Now, we will go more in-depth and focus on understanding some key parts of a cash flow statement.

There are three sections in a cash flow statement: cash flow from operating activities, investing activities, and financing activities. Cash flow from operating activities shows the *actual* amount of cash generated from business operations (making and selling products and services). Cash flow from investing activities shows the change in a company's cash due to purchases of long term assets. Cash flow from financing activities shows the cash gain or loss due to repaying debt, issuing stock, taking on debt, paying dividends, and similar types of charges.

Breakdown	TTM
˅ Cash flows from operating activ...	
Net Income	18,485,000
Depreciation & amortization	5,741,000
Deferred income taxes	-37,000
Stock based compensation	4,836,000
Change in working capital	7,250,000
Accounts receivable	-1,961,000
Accounts Payable	113,000
Other working capital	21,212,000
Other non-cash items	39,000
Net cash provided by operatin...	36,314,000

Above is part of Facebook's actual cash flow statement for 2019. I got the cash flow statement from Yahoo Finance. The section of the cash flow statement we are looking at is called "cash flow from operating activities."

The very top of the cash flow statement, which deals with cash flow from operating activities, starts with net income. The very first line of the cash flow statement, net income, will always be the exact same number as the last line on the income statement. The net income represents the profit a company made, and *it was calculated by following GAAP.* It represents the net income when depreciation and other non-cash charges are taken into account.

As explained earlier, revenues need to be recorded, or "matched," in the same period as the expenses that generate the revenue. In other words, expenses need to be recorded in the same accounting period (quarter or year) as the revenue that they generate. This is known as the GAAP matching principle.

For example, imagine that Facebook purchases desks and computers for their office, costing a total of $10,000. The desks and computers will help Facebook make money for many years into the future. For this example, let's say that they will use it for 5 years into the future, until it has to be replaced. According to GAAP matching principle, they cannot record $10,000 in expenses today. Rather, the expense has to be spread across over the items' useful life or life span of 5 years. If the cost is not split across the next 5 years, we would be violating the GAAP matching principle; if the expense is all recorded in one year, then the expense of the desks and computers would not be entirely matched with the time (5 years) that they are used to generate revenue for Facebook. According to GAAP, a $2000 expense would need to be recorded every year so that $10,000 worth of expenses can be recorded over 5 years, assuming a $0 salvage value and a straight-line method of depreciation for the purposes of this example.[25] *This process of expensing the cost of a long term asset into future accounting periods is known as depreciation in accounting.*

However, GAAP accounting's processes like depreciation that push revenues and expenses into future years can be troublesome for investors because it does not accurately portray a firm's ability to generate cash today. For example, imagine that a business does $100,000 per year in revenue, but purchased $1,000,000 worth of

---

[25] There are many methods used in accounting to calculate the depreciation expense. We are using straight line depreciation and assuming a salvage value of $0 because it makes the example simpler. It makes it easier to understand the basics of the cash flow statement. The purpose of this book is to help you learn investing, not accounting.

assets this year that will be used over the next 10 years. Their net income for the next 10 years would be $0. It would look like the business is not generating any cash for the owners because the depreciation expense of $100,000 per year would cancel out their revenue of $100,000, *assuming that they have no other expenses.* (According to GAAP's matching principle, they would need to expense the $1,000,000 over 10 years because they are planning to use the asset for 10 years, causing the business to have $100,000 in expenses per year while making only $100,000 in revenue). This would make the business look like it is performing very poorly.

However, taking a look at the same business from another perspective, one could say that the business invested $1,000,000, and it makes $100,000 back every year. It would then appear as if the company is getting 10% of their $1,000,000 investment every year. Most investors would consider this second portrayal, where the firm is getting a 10% return on their initial investment, as a more accurate portrayal of how well the business is performing. This is why we have the cash flow statement. The cash flow statement takes a look at businesses like in the second portrayal. It tracks exactly how much cash is entering and leaving a business, which can tell us about a stock's cash-generating ability.

Breakdown	TTM
˅ Cash flows from operating activ...	
Net Income	18,485,000
Depreciation & amortization	5,741,000
Deferred income taxes	-37,000
Stock based compensation	4,836,000
Change in working capital	7,250,000
Accounts receivable	-1,961,000
Accounts Payable	113,000
Other working capital	21,212,000
Other non-cash items	39,000
Net cash provided by operatin...	36,314,000

The cash flow statement takes the net income value which includes non-cash charges and makes adjustments to it so that it no longer reflects non-cash charges. For example, on Facebook's cash flow statement, the line items are being added/subtracted to net income in order to calculate the cash from operating activities.[26] For example, we see that the depreciation of $5.7 billion is being added to the firm's net income of $18.4 billion dollars. On Facebook's income statement, $5.7 billion would have been subtracted from revenue in order to calculate net income, even though no cash actually left the business. By adding the $5.7 billion with the net income on the cash flow statement, we are undoing the effect depreciation had on net income, giving us a more

---

[26] The word "line item" simply refers to each row on a financial statement.

accurate representation of how much cash the business actually generated.

Similar adjustments are made with the charges listed below depreciation like deferred income taxes, stock-based compensation, accounts receivable, and accounts payable (which are other types of non-cash charges). We are taking the firm's net income as reported on the income statement and undoing the non-cash charges to more accurately calculate how much cash actually entered the business. After making those adjustments, we are left with a number that more closely reflects how much cash entered or left the business *from core operations*. According to Facebook's cash flow statement, they generated about $36.3 billion from core operating activities.

The $36.3 billion in cash flow from operating activities does not represent the net amount of cash that Facebook generated. Cash from operating activities only includes expenses related to the core act of buying and selling goods and providing their service. Cash flow from operating activities does not include expenses related to long term reinvestment in the company. In order to **truly** calculate how much cash entered or left a business (free cash flow), we need to subtract capital expenditures (money spent on long term investment) from cash flow from operating activities.

Capital Expenditures

Capital Expenditures, also known as "CapEx," is usually found under the "cash flow from investing activities" section of a cash flow statement. On Yahoo Finance's cash flow statements, it is found on the second-to-last line on the cash flow statement. Capital expenditures are monies spent on long term assets in order to either grow or maintain the firm's current ability to generate cash.[27] It represents the amount of cash spent on new property, equipment, and other long term investments (expenses which are not included when calculating cash flow from operating activities).

When we take a firm's operating cash flow (which represents the net amount of cash that the business generated from selling products and services) and subtract capital expenditures, (cash spent on long term investments in the business) we are left with the net amount of cash that the business generated. This metric is called free cash flow.

Free cash flow is literally the amount of excess cash that the business produced. Free cash flow is the equivalent of taking the rent from a real estate property and subtracting all the money spent on running the property and reinvesting into the real estate business. It is the net cash that the property generated!

If you owned the entire company, you could do whatever you wanted with the free cash flow. You could buy a new car, pay off debt, or save it. It is all yours! However, when you are investing in stocks, the management of publicly traded companies makes the decision for you. They often use it to pay off some company debt, pay some dividends, and to form a stockpile of cash for future ventures.

Characteristic #4: of a Wonderful Business: It has Predictable Free Cash Flow

Free cash flow is one of the most important metrics to look at when deciding whether to invest in a business. It accurately represents how

---

[27] A more common term for "long-term assets" is "property, plant, and equipment."

much cash the business generates. If you know how much cash your property, business, or investment will generate, then you can determine how much to pay for it in order to get a good return on investment. This metric will be key in the next chapter when we calculate the intrinsic value of a stock. But for now, here is what you should look for.

On the cash flow statement, you want to see an upward trend in free cash flow. The reason is pretty obvious. A business that is growing its free cash flows is generating more cash each year. We want all of our assets to be generating more cash each year.

What is even more important than historical free cash flow growth is that the company's free cash flows will be predictable in the future. You want a company's free cash flow to be mostly in control of the company itself. In other words, you don't want to invest in businesses where their free cash flow is heavily dependent on outside, uncontrollable factors. Then, it is just hard to have a good estimate of future free cash flows.

The most common type of businesses that do have control of its free cash flows are businesses which are dependent on commodities.[28] Since commodities are standardized, a business will buy a commodity from whatever business offers it at the cheapest price. What is the point in paying more for the exact same good? In effect, the price of commodities is heavily dependent on supply and demand. If the commodity's supply goes up, the price goes down, and the opposite holds true. If the supply goes down, the price goes up. If commodity's demand increases, the price goes up, and once again, the opposite holds true.

---

[28]A commodity is simply a *uniform* good that is used in the production of other goods. They are *uniform*, so they are standardized goods. In other words, a commodity that you buy from one business will essentially be the same if you buy it from another business.

You never want to be invested in a business that makes a majority of their revenue from commodities. No one, including yourself, can predict the future supply and demand of commodities. There are just too many factors to take into consideration. There are also so many unpredictable factors which will affect their price in the future. There are unknown political factors which can change the supply or demand of a commodity. New, unknown technology and reserves can completely change the supply and price of a commodity. New and unknown consumer trends can completely change the demand of a commodity. Since you cannot predict the supply, demand, or price of any commodity, you cannot predict the revenues of a business that gets their revenue from selling a commodity. In essence, their cash-generating ability and free cash flow are not predictable.

If it is not possible to have a rough idea of a company's future cash-generating ability, then you can't determine how much to pay for the business. It would be like buying real estate without having an estimate of how much cash would be left after expenses. Investing in that would be just like gambling because you are hoping for a return. The same goes for businesses with unpredictable future free cash flows. You are just hoping that all will go well. If a business does not have predictable free cash flows, then you are just hoping for a good outcome. When you make an investment, you have to be sure that you will make a good return. If you are just hoping in the stock market, you are just gambling. So, you have to make sure that your businesses have predictable free cash flows.

Here are some types of businesses which don't have predictable free cash flows:

- The company is reliant on one or a few customers

- These businesses really don't have predictable cash flows over the long run because they are at the hands of another business. One day, their customer could say "I'm sorry, we don't want to work with you anymore." Then, all of a sudden, their free cash flows turn negative. Since you can't guarantee that a customer will still be a customer in a few years, you really can't get a good estimate of a company's future free cash flows if they only have a few customers. You just don't know whether and when a major customer will leave, so you can't estimate such a company's future free cash flows and hope to be accurate. This mainly occurs with B2B businesses.[29] In most annual reports, the company will have a breakdown of sources of revenue. Make sure their revenue sources are diversified.

- Free cash flow is currently not being generated by the firm

    - When a company is not generating free cash flows, it is very difficult for me to invest in the company. With these companies, I am usually not sure when the company will start generating money. Also, I usually do not have an idea of **how much cash** the company will be generating if it will be making free cash flow. In most cases, you just don't know when the business will start making money because you don't know when the management will physically make the necessary

---

[29] The phrase "B2B" simply means "business to business." It refers to a company that sells their product or service to another business. A B2C business sells their product or service to consumers.

changes so that the business can become profitable. You also don't know how profitable the business will be because you don't know how effective the management's actions will be. Since the future profits of businesses that don't make money today are very uncertain, it is best to avoid companies that don't generate cash, unless they are going to be profitable in the foreseeable future. Oftentimes, these companies are going to be called "the next big thing."

- As explained earlier, the company is dependent on commodities to make money (i.e. oil stocks)

    - No one can predict the amount of cash that a commodity-based business will generate over the long term. It is just not a good idea to invest in them because you can't determine a good price to pay for a business if you don't have an idea of how much cash they will generate.

- Businesses with inconsistent revenues

    - You always want to make sure that your stocks' revenues are consistent (i.e. not volatile) and trending upwards. If you are investing in a business where its revenues fluctuate a lot, it is an indicator that the business does not have predictable free cash flows. How can a company generate a predictable amount of cash without having predictable revenues? This is why it is so important to look at the income statement. Ensuring that your stocks' revenues are consistent and upward trending indicates that your businesses have predictable free cash flows.

- Businesses without moats and companies in rapidly changing industries

    - Two characteristics of a wonderful business are that they have a moat and are not in a rapidly changing industry. One of the big reasons why these characteristics make a business wonderful is because they make free cash flow very predictable. Strong moats and slowly changing industries prevent the competition from siphoning free cash flow. They also help prevent new competition from being successful, making wonderful business' free cash flow predictable.

Step 1 is Finding a Wonderful Business: Contract

Half of being successful at investing is being able to invest in wonderful businesses. Investing in these businesses will make your stocks compound dramatically, even if you did not buy them at a perfect price. So, it is very important to agree to the contract:

I _____ (insert your name) agree to invest in businesses that have the characteristics of a wonderful business:

- The stock is easily a better business than its competitors
- It has one of the three types of moats, whether it be owning a strong brand, being designated company in an area, or having competitive intellectual property.
- The industry is not rapidly changing
- It has a profit margin above 5%, a ROIC above 10%, a debt to asset ratio under .35, and a long-term debt to free cash flow ratio under 3.
- It has predictable free cash flows

Step 2 is Determining a Wonderful Price: Summary

Once you have found a wonderful business, you need to buy it at a wonderful price. A wonderful price is simply a good price to pay for the cash that a business will generate. It's like trading money. You give up money today for an asset that will produce cash tomorrow. In order to determine the wonderful price of a business, we need to determine how much a business is worth for the cash that it will generate. But we have to take into consideration that the cash that a business will produce in 2 years is not as valuable as the money that will be produced this year. So, we are going to need to find the *present value* of all the cash that a business will produce in the future, after we use a process to estimate the future free cash flows of a business.

Topics in this chapter include:

- The 2nd step of becoming a successful stock market investor
- A formula to calculate the wonderful price to buy a stock
- The ultimate method to predict the future free cash flows of a stock
- How time affects the value of money

Step 2 is Determining a Wonderful Price

"Intrinsic value can be defined simply: it is the discounted value of the cash that can be taken out of a business during its remaining life." -Warren Buffett

The intrinsic value of a stock is the ideal price to purchase a stock in order to get a good return on investment, and it is one of the most important concepts in this book. This chapter is going to outlay exactly how to calculate the intrinsic value or the ideal price to purchase a stock.

But before we look at an actual example of calculating the intrinsic value of a stock, we are going to look at an example of a simple, hypothetical business so that we can understand the overarching concept of intrinsic value.

Imagine that there is a business that is only going to operate for 2 years. And let's assume that you know that it is going to provide $100 in free cash flow per year; it has $100 leftover every year after paying expenses to run the business and for long term assets.

I want you to read this question and think about it before you continue reading: what is the value of the business if it will generate $100 per year *for 2 years*?

If I asked this question to a group of people, the most common answer I would get is "I would be willing to pay $200 for this business." Logically this makes sense. If a business is going to generate $200 in free cash flow over the next 2 years, shouldn't the business be worth $200?

However, this is not a totally correct statement to make due to a concept known as the time value of money. In order to understand the time value of money, imagine that you work at a very abnormal company and that they are very tight on cash. They do not have enough money to give many people their paycheck. Your paycheck is supposed to be $2,000 for this pay period. Your boss is looking for people who would be willing to take their paycheck at a later date. In order to get as many people to accept a late paycheck, your boss agrees to increase anyone's paycheck to $2,500, as long as they agree to receive their paycheck in 5 years (for simplicity's sake, let's just say that no one at this company was going to get paid more than $2,500 during this pay period). You would essentially be receiving a free $500 if you wait 5 years to receive your paycheck, while someone whose paycheck is currently $1000 would be getting a $1,500 bonus if they wait 5 years to receive it. Most people who are in your position and financially responsible would be patient and take the free $500. Intuitively, this sounds like a good decision because you would essentially get some free money.

However, it would be smarter to **not** be patient. Instead, you should take your paycheck today because money has a "time value." The sooner that you have cash, the more valuable it is. It is more valuable if you have it today because it has more time to compound and grow.

Let's say that you did not wait and took your paycheck on time, unlike most people. Then, you decided to invest it in the stock market, and you get a 10% return per year. That $2,000 would be worth $3,220.98 in 5 years, before any sort of fees, taxes, and inflation. Notice how that $3,220.98 is significantly more than the $2,500. When deciding whether to take $2,000 today or $2,500 in 5 years, *you have to convert the $2,000 into what it would be roughly worth in 5 years.* Then you are comparing apples to apples. Essentially, having $2,000 today is the equivalent of having $3,220.98 in 5 years. Would you rather receive a $2,5000 paycheck in 5 years, or have $3,220.98 in 5 years? I would take the $3,220.98 any day.

Since you are aware of the time value of money, you tell your boss that you will take your paycheck today. Your boss reluctantly agrees. However, you know that many of your colleagues are planning to take their paycheck late and to take the $2,500 because it is intuitively logical. They are not aware of the time value of money.

So, you decide to help them determine whether they should take their paychecks now or later. You have 3 colleagues, and the amounts that their paychecks are worth are listed below.

Colleague	Value of Paycheck (i.e. how much money they could receive today)
Colleague A	$1700
Colleague B	$1300
Colleague C	$1000

All of your colleagues have the option to take $2,500 in 5 years or to take their corresponding paycheck today. In order to decide whether it would be smart for your colleagues to take their paycheck now or later, you *could* follow the same process as how we determined that you should not be patient. However, that can be a time-consuming process because you would have to do the same calculation for each of your colleagues. So, instead of individually calculating what your colleagues' paychecks will be worth in 5 years and comparing their *future value* to $2,500, let's calculate the *present value* of having $2,500 in 5 years.[30]

In order to help your colleagues make a good decision, you can divide $2,500 by (1.1)^5 ($\frac{\$2,500}{(1+.1)^5}$). This will give us the *present value* of having $2,500 in 5 years. Remember that money has a time value and that it is more valuable if you have it sooner. In this case, your colleagues would have to wait in order to receive the $2,500. So, we need to calculate its true or *present* value, which we can calculate by dividing $2,500 by $(1+.1)^5$.[31] We are dividing by $(1+.1)^5$ because money has the ability to grow, on average, by about 10% per year in the stock market[32]. If money only had the capacity to grow by an average 5% per year, $(1+.05)^5$ would be in the denominator. Also, $(1+.1)$ is raised to the fifth power because the colleagues have to wait 5 years in order to receive the $2,500. The longer someone has to wait, the power becomes larger, the denominator becomes larger, and the present value becomes smaller (i.e. money is less valuable if you have to wait longer for it).

---

[30] Future value is a very important term in finance and investing. Future value refers to the value of an amount of money in the future. If you have money today, how much will it be worth in the future after it grows? *Present value* refers to how much money a future cash flow is worth in today's dollars, after we account for the time value of money. If we are going to receive $X in X years, what amount of money would we need today so that it can realistically grow to be worth $X in X years? **It is how much we would be willing to pay for a future cash flow (this will be explained later).**

[31] (1+.1)^5 can be rewritten as 1.1^5. For some people, it is easier to understand the expression when it is written as (1+.1)^5.

[32] 10% is only an average. It is very realistic to get much higher returns. Also, this example ignores taxes, fees, and inflation in order to keep it simple.

When you divide $2,500 by (1+.1)^5, we get a present value of $1,552.30. **This means that having $2,500 in 5 years has the equivalent value of having $1,552.30 today; $1,552.30 could grow to about $2,500 in 5 years.**[33]

Colleague	Value of Paycheck (i.e. how much money they could receive today)
Colleague A	$1700
Colleague B	$1300
Colleague C	$1000

Essentially, the boss is offering your colleagues two options. They can either take their paycheck as normal, with the amounts listed in the table, or they can take the boss's offer and take $2,500 in 5 years, **which is the equivalent of having $1,552.30 today and investing it.**

Now we can compare the offers in an apples-to-apples comparison, and it is much easier to see which choice is better for each of the colleagues. Colleague A can either take the $1,700 paycheck or take $2,500 in 5 years, **which is the same as having $1,552.30 today and investing it.** Colleague A should keep the $1,700 paycheck because it is worth more than $1,552.30. In other words, $1,700 can grow to an amount larger than $2,500 in 5 years. Colleagues B and C have a

---

[33] A 10% return is only an average return over investing for many years. Mutual funds can have a 20% return during a certain year and a loss during another year. Having $1,552.30 today does not **guarantee** that someone will have $2,500 in 5 years.

different situation. Their paychecks are worth less than $1,552.30. They should take the boss's offer and the $2,500 in 5 years because $2,500 is worth about $1,552.30 today, while their paychecks are worth about $1,300 and $1,000 today. If the colleagues take their $1,300 and $1,000 paycheck, it would have not likely grown to $2,500 in 5 years.

When we were deciding whether it would be good for the colleagues to take their paycheck early or late, we were calculating how much money a person should be willing to give up in order to have $2,500 in 5 years.

Investing in the stock market, or in any financial asset, is very similar to the problem with the colleagues. The $2,500 is just like free cash flow for a business. An investor has to determine what is a good price to pay (i.e. how much cash to give up today) for the free cash flow that a business will generate in the future, like how we had to determine how much cash the colleagues should be willing to give up for the $2,500 in free cash flow. When solving the problem with the colleagues, we discounted the $2,500 free cash flow in order to determine its *present value*.[34] We can use the same process and calculations when determining the value of a business; we can calculate the present value of all of its future free cash flows. This will tell us the value of all the cash that a business will produce in the future.

This process of discounting future free cash flows is used by so many of the best investors. Bill Ackman, Warren Buffett, and Charlie Munger, and Seth Klarman all agree that the intrinsic value of a stock is the present value of all its future cash flows. In fact, during a 1998 annual meeting, Warren Buffett said the following:

---

[34] The word "discounted" just refers to the process of calculating the present value of a future free cash flow.

"In order to calculate intrinsic value, you take those cash flows that you expect to be generated and you discount them back to their present value."

In this chapter, you are going to learn the process of discounting free cash flows. You are going to learn how to value a business and determine a good price to buy a stock. But first, let's value a very simple, hypothetical business before we get to discounting an actual stock.

Let's say that you found a business which you *know* is going to generate about $100 in free cash flow next year and close in a year's time. Let's assume that you want to get at least a 10% return on investment because you want to outperform the mutual fund managers of Wall Street.

In order to calculate how much you would need to pay for the business in order to get a 10% return on investment, you would use the following formula:

Intrinsic Value = $Free\ Cash\ Flow/(1+r)^t$

t = time until the free cash is received
r = rate of return that we want our money to get (i.e. the required rate of return)

In order to calculate the intrinsic value of this business, you would need to divide $100 (free cash flow) by 1.1 (a 10% rate of return). The intrinsic value of this business is $90.90 because buying it at $90.90 will allow you to get the desired rate of return. Investing in a business or stocks is like trading money. You can give up $90.90 for an asset which will produce $100 for you next year, allowing you to get a 10% return.

This example was fairly simplistic because most businesses generate cash for many years. How do we calculate the intrinsic value of a business which is going to generate cash for multiple years?

Let's calculate the intrinsic value of the same business, but this time, let's assume that the business will be operating for 2 years. This time, we are going to organize our work on a table because it gets a little more complex when there are multiple years involved.

	Year 1	Year 2
Free Cash Flow	$100	$100
Discount Factor	$(1+.1)^1$	$(1+.1)^2$

The first row of the table just shows the free cash flow that the business will produce over the next 2 years. The second row is known as the discount factor. The discount factor is a fancy term in finance. It simply refers to the number in the denominator when we are calculating the intrinsic value of a stock's future free cash flows. In this book, we will not be specifically using the term "discount factor" very often, but it is good to know that it is formally called the discount factor. In this book, we are going to be calling the discount factor as "the required rate of return."

Let's think about the free cash flow and discount factor for each year separately. The business is going to produce $100 next year (year 1). In order to get a 10% return on investment *for the cash that is going to be produced next year*, you would need to pay $90.90 today. It is the same amount as the last example; you give up $90.90 today for $100 next year to get a 10% return.

During year 2, the business is going to produce another $100. The fact that you would have to wait 2 years to receive it is important.

Remember that money is more valuable the sooner that you have it because it can grow over time through investing. Conversely, money is less valuable the later that you receive it. The same applies to a business. If it produces cash later, it is less valuable.

When we are calculating the discount factor for year 2, you need to keep in mind that it will take the business 2 years to produce that free cash flow. So, we would raise 1.1 to the second power this time. Dividing $100 by 1.1 to the second power gives us the intrinsic value of the firm's free cash flow during year 2, which is $82.64. In other words, paying $82.64 for $100 in 2 years would give a 10% return on investment *per year* ($82.64 could grow to about $100 in 2 years, assuming a 10% return per year).

	Year 1	Year 2
Free Cash Flow	$100	$100
Discount Factor	1.1^1	1.1^2
Present Value of Free Cash Flow	$90.90	$82.64

So, we have a business whose free cash flows have a present value of $90.90 and $82.64. What is the intrinsic value of the entire business?

In order to calculate the present or intrinsic value of the entire business, we need to take the sum of the present values. In this case, we would take the sum of $90.90 and $82.64. $90.90 and $82.64 are the present values of the firm's future free cash flows for each year individually, so taking their sum will give you the present value of all the firm's future free cash flows. The sum, $173.54, represents the intrinsic value or *net present value* of the business (i.e. giving up $173.54 today will allow us to get a 10% return per year if we invest in a business that produces $100 next year and $100 the year after that).[35]

Here is an equation of that represents the calculation that we just solved:

$$Intrinsic\ value = \frac{FCF1}{(1+r)^1} + \frac{FCF2}{(1+r)^2}$$

$$Intrinsic\ value = \frac{\$100}{(1+.1)^1} + \frac{\$100}{(1+.1)^2} = \$173.54$$

In the last example, we assumed that the business was only going to operate for 2 years. In reality though, businesses operate for many years, and we are hoping to invest in businesses which will last for many decades. So, we need a formula that will allow us to value a business that we expect to operate over the long term.

Here is a formula that will allow us to calculate the intrinsic value of a company that we can expect to operate in perpetuity. This formula allows us to calculate the exact present value of future free cash flows produced over a company's lifetime.

$$Intrinsic\ value = \frac{FCF1}{(1+r)^1} + \frac{FCF2}{(1+r)^2} + \frac{FCF3}{(1+r)^3} + \frac{FCF4}{(1+r)^4} + \frac{FCF4(1+g)}{r-g} * \frac{1}{(1+r)^4}$$ [36]

---

[35] The *net present value* of a business is simply a term in finance and investing which refers to the sum of the *present values* future cash flows. In this book, net present value is also a synonym for intrinsic value.

[36] I just want to make it clear that the number after the words "FCF" are not a constant. For example, FCF2 simply refers to the free cash flows which will be generated in 2 years from now. FCF2 does not mean to multiply a future free cash flow by 2.

FCF1= Free cash flow in 1 year
FCF2= Free cash flow in 2 years
FCF3= Free cash flow in 3 years
FCF4= Free cash flow in 4 years
r= Required rate of return (10%)
g= Perpetual growth rate of the firm

The first 4 terms of the equation are pretty straightforward. They are simply discounting the firm's free cash flows over the next 4 years and calculating their present value. The last term is new, and its concept was not in the other examples.

The last term is known as the terminal value. The terminal value is represented by the following equation: $\frac{FCF4(1+g)}{r-g} * \frac{1}{(1+r)^4}$. When we plug in the variables, it will tell us the present value of the all firm's free cash flows after year 4. When we take the sum of the present values of the firm's free cash flows after 4 years and for the first 4 years, we get the intrinsic value of the company; we get the present value of all the money that the company will generate.

Let's define all of the terms in the equation before moving on. The "g" is known as the perpetual growth rate. It is the average annual growth rate that we can safely say that the company will continue to grow its free cash flows in perpetuity. We will look at how to derive this number once we get into our example of calculating the intrinsic value of a real stock. The FCF4*(1+g) represents the firm's free cash flow estimate in year 5. In other words, it is the free cash flow of year 4, multiplied by the perpetual growth rate to estimate the free cash flow of year 5. The "r" value is the same number that we have been using in all of the examples; it is the 10% required rate of return.

Now, let's apply the terminal value in order to value a hypothetical business.

Let's say that a firm is going to make $100 in free cash flow per year over the next 4 years, and after that, its free cash flows are going to grow at about 2.5% per year in perpetuity. In this scenario, we are going to assume that our required rate of return is still 10%.

$$\text{Intrinsic value} = \frac{FCF1}{(1+r)^1} + \frac{FCF2}{(1+r)^2} + \frac{FCF3}{(1+r)^3} + \frac{FCF4}{(1+r)^4} + \frac{FCF4(1+g)}{r-g} * \frac{1}{(1+r)^4}$$

Since we now understand the formula, and since we know the values of the variables, we can simply plug and chug in order to calculate the intrinsic value of the company.

$$\text{Intrinsic value} = \frac{100}{(1+.1)^1} + \frac{100}{(1+.1)^2} + \frac{100}{(1+.1)^3} + \frac{100}{(1+.1)^4} + \frac{100(1+.025)}{.1-.025} * \frac{1}{(1+.1)^4} = \$1250.43$$

When we plug in the numbers, we get the true intrinsic value of the business. A business that makes $100 per year in cash flow for 4 years then grows its cash flow by 2.5% per year in perpetuity is worth $1250.43, if the required rate of return is 10%. However, it is important to note that the calculation likely seemed simple right now. That is because the future free cash flows were given. When we go into calculating the intrinsic value of an actual stock, we will go over how to determine the firm's future free cash flows.

Right before we go into an example with a real stock, it is important that we take one last look at the terminal value calculation. Many people, when they first see the calculation for terminal value, get confused because the mathematics does not make sense. How can the expression ($\frac{100(1+.025)}{.1-.025} * \frac{1}{(1+.1)^4}$) represent the present value of the firm's future cash flows in perpetuity? In order to prove the accuracy of the formula, I created a spreadsheet that calculated the intrinsic value of the same business.

The spreadsheet has 3 columns. The first column is the year. The second column is the free cash flow the firm is going to produce in a given year. The value of the free cash flow is simply $100 for the first 4 years, and $100 with a 2.5% growth rate for every year in perpetuity after the 4th year. In the third column, I calculated the present value of the free cash flow *produced during that year*. I then took the sum of the 3rd column. This would give me the intrinsic value of a company or the present value of all the free cash flows that the business will produce. When I added up the 3rd column of the table, I got $1250.43. Notice that $1250.43 is the same number that we got when we calculated the intrinsic value of the same business using the equation. It proves that the formula is equivalent to the intrinsic value of a stock: the present value of all the free cash flows that a business will produce.

Year	Free Cash Flow	Present Value of Cash Flow $(\frac{Free\ Cash\ Flow}{1.1^{Year}})$
1	100	90.909
2	100	82.644
3	100	75.131
4	100	68.301
5	102.5	63.644
6	105.06	59.305
7	107.68	55.261
8	110.38	51.493
9	113.14	47.982

10	115.96	44.711
11	118.86	41.662
12	121.84	38.822
13	124.88	36.175
14	128	33.708
15	131.2	31.41
16	134.48	29.268
17	137.85	27.273
18	141.29	25.413
19	144.82	23.68
20	148.45	22.066
21	152.16	20.561
22	155.96	19.159
23	159.86	17.853
24	163.86	16.636

*As a side note, I only copied the first 24 years of my spreadsheet into this book to save paper.[37]

---

[37] It could sound weird that we are assuming that a stock will be operating in perpetuity. If we are assuming that it will be producing cash forever, won't we be overestimating the true intrinsic value of a business because all business don't last forever? Well, we actually won't be overestimating it, and here is why. A characteristic of a wonderful business is that it will operate for many years and decades into the future. Cash that is produced many years into the future has a very insignificant present value. For example, the present value of $100 that is produced in 50 years is worth less than $1. So, if you expect a

Calculating the Intrinsic Value of a Real Stock

When we were calculating the intrinsic value of the hypothetical businesses, I told you the free cash flows which the business was going to produce each year. This information allowed you to calculate the present value or intrinsic value of those cash flows. In the world of investing, you are not told the future free cash flows which a business will produce. Instead, it is the job of an investor to predict the future free cash flows which a business will produce. So, the first step to calculating the intrinsic value is to estimate future free cash flows.

The challenge that we now face is predicting future free cash flows. Luckily, there is a multi-step process that we can follow in order to estimate the future free cash flows of most businesses. The first step to estimate future free cash flows is to estimate a company's future revenue. We will then use the revenue predictions in order to estimate their net income, which will help us estimate free cash flow because the two metrics correlate with each other; if net income rises, free cash flow tends to rise.

---

business to produce $100 per year for 100 years and another business to produce $100 per year for 1000 years, they would have very similar intrinsic values because cash that is produced in years 101-1000 has very small and insignificant present value. In essence, the cash produced in years 101-1000 would add very little to the intrinsic value of the business. So, if you calculate the intrinsic value of a business and invest in it assuming that it will operate in perpetuity, but it really only survives for 50 years, you would have still get a good return on investment because the cash you expected to be produced after 50 years would have added very little to the intrinsic value of the stock and its ideal price to buy it.

There are multiple steps in the process of estimating future free cash flows. In order to keep your work organized, I suggest that you create a table, as you will see in upcoming pages. It will prevent you from losing track of your work in the middle of your calculations. Now, let's get into estimating the future free cash flows of a real stock: Best Buy.

First, We Need to Estimate the Stock's Future Revenue

Our first step for predicting the future free cash flows of a business is going to be to estimate the company's future revenue. Conveniently, most stock research websites provide revenue estimates. We can use those estimates in order to predict the firm's future free cash flows. I inserted Best Buy's estimated revenue in the table below. I found them on Yahoo Finance and under the analysis tab. However, most stock research websites only have predictions for the next 2 years, and we want estimates for the next 4 years for our calculations; forecasting revenue for additional years makes our calculations more accurate as long as our estimates are realistic. In order to forecast a stock's revenue for the subsequent 2 years (i.e. years 3 and 4), we can look at how revenue is expected to trend and how it has been trending.

	Historical Performance				Estimations			
Year Ending Feb.	2017	2018	2019	2020	2021	2022	2023	2024
Revenue	39.403 Billion	42.151 Billion	42.879 Billion	43.638 Billion	43.6 Billion	44.5 Billion		
Revenue Growth	-0.31%	6.90%	1.70%	1.70%	0.09%	2.00%		

So, I looked at Best Buy's income statements. Then in the table above, I added Best Buy's revenue for 2017 through 2020.[38] Then I calculated the revenue growth rate for each of those years. I also calculated the revenue growth rate for 2021 and 2022.[39]

We can use the company's historical and predicted growth to estimate their revenue for 2023 and 2024. Since 2017 and going into 2022, the average of all the revenue growth rates is 2.013% ($\frac{-.31\% + 6.9\% + 1.7\% + 1.7\% + 0.09\% + 2\%}{6}$). Therefore, it would be safe to assume that Best Buy is going to grow its revenues by about 2% per year over the next few years, including 2023 and 2024. We can use the 2% growth rate that we just calculated in order to predict their revenues for 2023 and 2024. To predict their revenue for 2023, we can simply multiply their expected revenue for 2022 by 2% ($44.5 * 1.02= $45.39). Similarly, to predict their revenue during 2024, we can multiply their expected revenue during 2023 by 2% (45.39* 1.02= $46.297) Then, we just need to enter those values in the table.

It is important that I did not grow their revenue by 2.013%. When estimating the intrinsic value of any business, it is always good to use more conservative numbers (i.e. inputs that will make the intrinsic value lower).[40] This adds a slight margin of safety into the calculation,

---

[38] If you are reading this book in 2020, you might be wondering how I know Best Buy's revenue for 2020 if the year has not ended yet. Well, Best Buy's fiscal year ends in February, so fiscal year 2020 ended in February of 2020 for Best Buy.

[39] To calculate a growth rate between two adjacent years, you can use the following formula: (The change in revenue (or any other metric) between 2 years)/(the value of the metric in the initial year)

[40] In investing and business, the word "conservative" refers to the fact you should be less optimistic about your estimates. This ensures that if something

just in case the business growth is less than expected. This will ensure that we are not overpaying for a business, even if things don't go as planned. Although this margin of safety is very small, it illustrates an important habit to have when calculating intrinsic value: always round down. Rounding down throughout the calculations can help create a margin of safety.

Next, We Need to Estimate the Stock's Profit Margin and Net Income

The next step is to estimate what the stock's profit margin will be in the future. Except in certain cases, this is not a volatile figure for most businesses; the profit margin hovers around the same number every year. Businesses that don't have a stable profit margin usually have very volatile revenues and unpredictable free cash flows. Since profit margins are not volatile, we can estimate a stock's future profit margin by taking an average of their historical profit margins.

Their historical profit margins can be calculated by looking at their income statement and dividing their net income by their total revenue ($\frac{Net\ Income}{Total\ Revenue}$).For new investors, net income can be found at the bottom of an income statement, while total revenue can be found at the top. I have done these calculations for Best Buy's historical performance. Their historical profit margins are listed in the table that is below.

---

goes wrong in your estimates, the situation is not as bad. Being conservative is similar to taking a margin of safety.

	Historical Performance				Estimations			
Year Ending Feb	2017	2018	2019	2020	2021	2022	2023	2024
Revenue	39.403 Billion	42.151 Billion	42.879 Billion	43.638 Billion	43.6 Billion	44.5 Billion	45.39 Billion	46.297 Billion
Revenue Growth	-0.31%	6.90%	1.70%	1.70%	0.09%	2.00%	2%	2%
Profit Margin	3.11%	2.37%	3.41%	3.53%	3%	3%	3%	3%

Best Buy's average profit margin from 2017 through 2020 has been about 3.1%. Since their profit margin is about 3.1%, we can safely assume that it will be at least 3% from 2021 through 2024.[41] Notice how I took a more conservative number again. I'm going to assume that they are going to only have a profit margin of 3% instead of 3.1%. By estimating that they are going to be a little less profitable, we are adding a little more safety in our calculation because we will end up with a lower intrinsic value. Just in case profits temporarily fall due to a short-term issue, the margin of safety helps prevent us from paying for cash flows which will never be generated.

The next step is fairly simple. In the table, we need to write down Best Buy's net income from 2017 through 2020. This is simply the same number that was used in the numerator to calculate the profit

---

[41] When taking the averages of the profit margin to estimate a future profit margin, it is important to remove any one-time outliers. These are usually caused by pieces of news and by large, one-time purchases which cause profits to temporarily fall and quickly recover. If an event like this is not expected to happen to the business in the future, then the profit margin for an outlier year should not be used to estimate a future profit margin, especially if it is significantly different from its adjacent years.

margin. Once again, net income can be found on the last line of the income statement.

Next, we need to estimate Best Buy's net income for future years. In order to get an estimation of their net income, we have to multiply their estimated profit margin by their expected revenue for each year. Since a profit margin just represents the portion of revenue that becomes profit, multiplying the firm's expected revenue by the expected profit margin will give us a good estimate of their future net income. For example, to estimate the firm's profit during 2022, we would multiply $44.5 billion by 3%, giving us $1.335 billion in expected net income during 2022. Similarly, to predict the firm's profit during 2024, we can multiply their expected revenue of $46.297 billion by their profit margin of 3%, giving us an estimate of 1.388 billion in net income for 2024.

	Historical Performance				Estimations			
Year Ending Feb	2017	2018	2019	2020	2021	2022	2023	2024
Revenue	39.403 Billion	42.151 Billion	42.879 Billion	43.638 Billion	43.6 Billion	44.5 Billion	45.39 Billion	46.297 Billion
Revenue Growth	-0.31%	6.90%	1.70%	1.70%	0.09%	2.00%	2%	2%
Profit Margin	3.11%	2.37%	3.41%	3.53%	3%	3%	3%	3%
Net Income	1.228 Billion	1 Billion	1.464 Billion	1.541 Billion	1.308 Billion	1.335 Billion	1.361 Billion	1.388 Billion

There are only a few more steps that we need to do in order to predict the firm's future cash flows.

Now, We Need to Estimate Future Free Cash Flows

The next step is to see how net income correlates with the firm's free cash flow. If we are able to derive a rule that says if a company has X amount of dollars in net income, they are likely to have about Y amount of dollars in free cash flow, we would be able to predict their future free cash flow. Since we have a good estimate of the firm's future net income, we would be able to predict the Best Buy's future free cash flows if we have a rule that estimates their free cash flow given their net income. Luckily, we can use a firm's history to derive such a rule.

In order to derive the rule, we first need to add the firm's historical free cash flows in our table. Most stock research websites will have data on their free cash flow for each year. On Yahoo Finance, it can be found at the bottom of the cash flow statement under the "Financials" tab. Copying it into the table will help us stay organized.

Then we need to divide the firm's free cash flow by their net income for each historical year. The ratio will show the amount of free cash flow the firm had for every dollar of net income. The ratios are listed in the next table.

If we take the average of the ratios, we can see that, on average, the firm produces $1.32 in free cash flow for every dollar of net income.[42]

The free cash flow to net income ratio will be around that average over the long run. Since we are investing in predictable businesses in slowly changing industries, the free cash flow to net

---

[42] When doing this calculation, you might find outliers that are a one-time occurrence. If this is the case, you should not include them in your calculation.

income ratio will not change much in future years. So it is very safe to assume that their free cash flow to net income ratio will be around 1.32 in future years.

However, since it is important to be conservative with our estimates when doing discounted cash flow analysis and calculating the intrinsic value, we are going to assume that the ratio is going to be about 1.1 in future years. This is about 16% lower than their four-year average. When deciding to use 1.1 instead of 1.32, I took into account that their more recent years have had a lower ratio. So, the ratio in the upcoming years will likely be slightly lower than the 4-year historical average. For example, when we take the average of their 2019 and 2020 free cash flow to net income ratios, we get an average of 1.13. When compared to 1.32, 1.1 is a more accurate and safe estimate based on their more recent financial performance.

However, if I personally felt that Best Buy was going to be having significantly stronger years in the future, I would have not used 1.1 as the ratio. It would be much better to take a smaller margin of safety if expectations are more optimistic and certain. If we saw that the free cash flow to net income ratio was consistently increasing, using the historical, average free cash flow to net income ratio would likely give us a slight underestimate of the firm's ratio in the future. Therefore, only a small margin of safety, if any at all, would be needed when the future is expected to be more optimistic. This is one reason why it is important to know and understand the business that you are investing in. You can end up taking too much or too little margin of safety if you are not sure of their future prospects.

Year Ending Feb	Historical Performance				Estimations			
	2017	2018	2019	2020	2021	2022	2023	2024
Revenue	39.403	42.151	42.879	43.638	43.6	44.5	45.39	46.297

	Billion	Billion	Billion	Billion	Billion	Billion	Billion	Billion
Revenue Growth	-0.31%	6.90%	1.70%	1.70%	0.09%	2.00%	2.00%	2.00%
Profit Margin	3.11%	2.37%	3.41%	3.53%	3%	3%	3%	3%
Net Income	1.228 Billion	1 Billion	1.464 Billion	1.541 Billion	1.308 Billion	1.335 Billion	1.361 Billion	1.388 Billion
Free Cash Flow	1.963 Billion	1.453 Billion	1.589 Billion	1.822 Billion	1.438 Billion	1.468 Billion	1.497 Billion	1.526 Billion
FCF to Net Income	1.59	1.453	1.08	1.18	1.1	1.1	1.1	1.1

Our next step is to multiply our conservative estimate of the free cash flow to net income ratio by our net income estimations. This will provide us with the firm's future free cash flows. For example, to find the firm's future free cash flows in 2021, we would multiply Best Buy's expected net income of 1.308 billion by 1.1. This will provide us a good estimate of the firm's future free cash flow during 2021. We can follow the same process for each of the other years. We can multiply 1.1, the average free cash flow to net income ratio, by the firm's expected net income. Doing this will provide a good, conservative estimate of the firm's free cash flow from 2021 through 2024.

Intrinsic value
$$= \frac{FCF1}{(1+r)^1} + \frac{FCF2}{(1+r)^2} + \frac{FCF3}{(1+r)^3} + \frac{FCF4}{(1+r)^4} + \frac{FCF4(1+g)}{r-g} * \frac{1}{(1+r)^4}$$

Looking back at our equation to calculate the intrinsic value of a company, we now know the numerators of the first 4 terms and part of the 5th numerator; we have a good estimate of the firm's free cash flow over the next 4 years. In our past calculations, we have been using 10% as the required rate of return. But, in the world of finance, there are many ways to calculate the "r" value. So, let's take a look at what is the ideal "r" value that you should be using.

Next, We Need to Input the "R" Value

The next step is going to be to determine the "r" value or the required rate of return. There are two ways to determine the required rate of return. One of the methods is commonly used in the world of finance and usually leads to an investor underperforming the market. The other method is preferred by intelligent investors like Warren Buffett.

In the world of finance, a common method used by some investors to calculate "r" is to look at a firm's weighted average cost of capital and use it as the "r" value. According to finance theory, investors should use the weighted average cost of capital as their required rate of return. The weighted average cost of capital shows the average cost of capital for the firm; it shows how much it costs to finance the company through debt and equity (i.e. the cost for the company to raise money). For example, a company that is only financed through debt would have a weighted average cost of capital that is equal to the interest rate of the loans. If a company is financed through debt and equity, the weighted average cost of capital would equal to the following formula:

$$Cost\ of\ Equity * \%\ Equity + Cost\ of\ Debt * \%\ Debt.$$

The equation says that the weighted average cost of capital is equal to the portion of the company that is financed through equity times the cost of equity, plus the portion of the company that is financed through debt times the cost of debt (i.e. the interest rate).

In real life, the weighted average cost of capital is commonly used to help management determine if a new project is worth pursuing. Imagine that a restaurant is looking to open a new location. In order to determine whether it would be a good investment, its management would look at the firm's weighted average cost of capital and compare it to the project's expected returns. Let's say that the restaurant sells $500,000 worth of bonds (i.e. debt) to raise capital and that the restaurant's weighted average cost of capital is 5%. In other words, the interest rate of that debt is 5%. The management would receive $500,000 in cash, but they would be obligated to pay $25,000 in interest payments. In order to be profitable, they would need to generate at least $25,000, or get a 5% return on their $500,000 investment.[43]

According to finance theory, the restaurant and all companies would only make investments that would yield them a return greater than the weighted average cost of capital. If the firm pursues a project that yields a return less than the weighted average cost of capital, they would not cover the costs of financing it. If the restaurant only thinks that it will make $24,000 in gross profit, the firm would not even be able to cover the interest payments. Then they would decide to not pursue the project.

In theory, since a company would not pursue ventures that would yield a return lower than the weighted average cost of capital, they would have a return on investment that is at least equal to the weighted average cost of capital. So, many investors use the weighted average cost of capital as the required rate of return or the discount rate.

---

[43] This example assumes that the company is not financed at all by equity. This makes it a little more simple to understand.

Although it can be logical **for a business** to use the weighted average cost of capital in order to determine whether a venture will be profitable, **as an investor**, you should not use the weighted average cost of capital as the required rate of return when investing in the stock market.

There is one **major flaw** with using the weighted average cost of capital as the required rate of return; the weighted average cost of capital does not reflect opportunity cost. Opportunity cost is an important concept in business. It is the opportunity that is lost from choosing one option over another. Being able to make good decisions, whether it be about stocks or anything else in life, comes down to being able to conduct a thought-out opportunity-cost analysis.

When I was at a car dealership last year, I saw someone making a decision that many people have to make. The person was looking at a very nice and new van. But he looked unsure. He was eyeing a used and cheaper car that was just a few feet away. The car was several thousand dollars cheaper than the van. But the van would make the man feel good because he really wanted to drive that van. He was faced with a difficult decision. Should he spend an extra couple thousand dollars in order to have a vehicle that he really wants, or should he save a few thousand dollars and get a car that he does not desire as much? For many people, this would be a tough call to make, while for others, it would be a no-brainer. Although I was unable to witness the final decision, whatever vehicle he chose, he was faced with a decision and had to consider the opportunity cost of choosing one vehicle over another. If he got the van, his opportunity cost was his potential to save money by purchasing the car. If he purchased the car, his opportunity cost would have been the loss of happiness from owning the van. He would need to weigh whether the additional cost of purchasing the van is worth the additional satisfaction of purchasing the van over the car. By considering what you give up when making a decision, you are considering the opportunity cost of it, which is key to making a good decision in life and investing.

Let's bring the concept of opportunity cost back to investing. Let's say that we know that this business will provide $100 in free cash flow in perpetuity. Also, let's assume that we calculated the company's weighted average cost of capital to be 7%.[44] If we use the 7% weighted average cost of capital as the required rate of return or discount factor, then we are essentially saying that we want to purchase the stock so that we get a 7% return on investment. Using our formula for intrinsic value we would have the following:

$$Intrinsic\ value = \frac{FCF1}{(1+r)^1} + \frac{FCF2}{(1+r)^2} + \frac{FCF3}{(1+r)^3} + \frac{FCF4}{(1+r)^4} + \frac{FCF4(1+g)}{r-g} * \frac{1}{(1+r)^4}$$

$$Intrinsic\ Value = \frac{\$100}{(1+.07)^1} + \frac{\$100}{(1+.07)^2} + \frac{\$100}{(1+.07)^3} + \frac{\$100}{(1+.07)^4} + \frac{\$100(1+0)}{(.07-0)} * \frac{1}{(1+.07)^4}$$

Using the equation, the intrinsic value of its cash flows is roughly $1428. Investors who use the weighted average cost of capital as the required rate of return would purchase this hypothetical business for about $1428. However, this decision is not great because it does not reflect opportunity cost. Instead of investing in that company which will provide us a 7% return, an investor could invest in an index fund or the S&P 500 and get an average return of 10% per year. Here, the opportunity cost is very clear. By choosing to get a 7% return by

---

[44] 7% would be a normal weighted average cost of capital for a company.

investing in the hypothetical business, an investor would be losing the opportunity to get a 10% return in an index fund. Why would someone want to invest in the business and get a 7% return when they can just get an easy 10% return through an index fund?

Instead of using a firm's weighted average cost of capital as the discount factor or required rate of return, you should use 10% as the discount factor. It will ensure that we are accounting for the fact that you have the option to invest in the S&P 500 and get an average return of 10% per year. By having the discount factor at 10% and by always investing in businesses that are trading below their intrinsic value, we are ensuring that we are purchasing businesses at a price so that the firm's future cash flows provide **at least** a 10% return per year. This will ensure that we are purchasing business so that our long-term returns are at least greater than that of the market.

So, we will use a required rate of return of 10% when calculating the intrinsic value of Best Buy in order to ensure that we are going to beat the market.

Looking at the following table and formula, we can see that we have almost all the data that we need in order to calculate the intrinsic value of Best Buy. We have a good estimate of the firm's free cash flow for the next 4 years, and we know that we should be using 10% as the required rate of return. Now, we just need to figure out the "g" value.

	Historical Performance				Estimations			
Year Ending Feb	2017	2018	2019	2020	2021	2022	2023	2024
Revenue	39.403 Billion	42.151 Billion	42.879 Billion	43.638 Billion	43.6 Billion	44.5 Billion	45.39 Billion	46.297 Billion
Revenue Growth	-0.31%	6.90%	1.70%	1.70%	0.09%	2.00%	2.00%	2.00%
Profit Margin	3.11%	2.37%	3.41%	3.53%	3%	3%	3%	3%

Net Income	1.228 Billion	1 Billion	1.464 Billion	1.541 Billion	1.308 Billion	1.335 Billion	1.361 Billion	1.388 Billion
Free Cash Flow	1.963 Billion	1.453 Billion	1.589 Billion	1.822 Billion	1.438 Billion	1.468 Billion	1.497 Billion	1.526 Billion
FCF to Net Income	1.59	1.453	1.08	1.18	1.1	1.1	1.1	1.1

$$Intrinsic\ value = \frac{FCF1}{(1+r)^1} + \frac{FCF2}{(1+r)^2} + \frac{FCF3}{(1+r)^3} + \frac{FCF4}{(1+r)^4} + \frac{FCF4(1+g)}{r-g} * \frac{1}{(1+r)^4}$$

Lastly, We Need to Determine an Appropriate Perpetual Growth Rate, "g"

The "g" value or perpetual growth rate represents the expected, average free cash flow growth rate over the entire life span of the business after the first four years that were forecasted. Determining a proper "g" value is crucial when calculating the intrinsic value of a business. The intrinsic value is very sensitive to the "g" value, so we need to make sure that we do not determine it to be too large, or else we will end up with an overly optimistic valuation.

Let's look at how we can determine the perpetual growth rate. In order to determine the "g" value, we can look at the firm's expected and historical free cash flow growth rate. In the table below, I have calculated the free cash flow growth rate for each year.

	Historical Performance				Estimations			
Year Ending Feb	2017	2018	2019	2020	2021	2022	2023	2024
Free Cash Flow	1.963 Billion	1.453 Billion	1.589 Billion	1.822 Billion	1.438 Billion	1.468 Billion	1.497 Billion	1.526 Billion
Free Cash Flow Growth Rate	184%	-25.9%	9.3%	14.6%	-21.0%	2.08%	1.9%	1.9%

Since Best Buy is a mature, slow-growth company, we can use the growth rates in the chart to predict the Best Buy's perpetual free cash flow growth rate.[45] In order to calculate the firm's expected free cash flow growth rate, we can look at the average of the historical and expected growth rates. **By taking the average of the firm's historical and estimated free cash flow growth rates, we can get a good estimation of their long-term growth rate or "g" value.** However, if you are analyzing a mature, slow-growth company and get an average free cash flow growth rate or "g" value between, 2%-3%, it is best to round down and use a "g" value of 2% because it allows us to account for the fact that a company's growth will slow down in the long run. If you get an average between 1% and 2%, use 1% as the "g" value because, once again, it allows you to account for the fact that a company's growth will slow down in the future. Similarly, if you get an average between 3%-

---

[45] In this book, I am considering a slow-growth company as a company that is expected to grow their free cash flows by less than 4% per year.

4%, use 3% as the "g" value. Use 0% if you get an average between 0% and 1%. Essentially, a good rule of thumb is to always round down your "g" value estimate.

It is important to note that we can take the average of the free cash flow growth rates and use it to estimate future cash-flow growth because Best Buy is a mature, slow-growth company with stable growth. If a company that has been **consistently** growing their free cash flows by more than 4%, historical growth would not reflect perpetual growth because the growth of all companies has to slow down to a more stable growth rate; such a fast growth rate is not sustainable.

For fast-growing companies, it is best to use a "g" value of about 3%-4%.[46] This essentially means that, after the 4th year, we are estimating that free cash flows will grow by about 3%-4% per year on average.[47]

For fast-growing companies, no one knows when the growth will slow down and how much it will slow down by. It is all a mystery until it does in fact slow down. So it is best to use a conservative estimate of 3%-4% growth per year for fast-growing companies. By assuming a 3%-4% growth rate after the 4th year, we are assuming that the company's growth will slow down significantly after the 4th year.

---

[46] **If a company has been growing free cash flows by 4%-10%, use a "g" value of 3%. If a company has been growing free cash flows by more than 10%, use a "g" value of 4%.**

[47] Some investment theorists and professors say that you should never use a "g" value larger than the GDP growth rate (about 2.5% during most years). They say this because "having a company growth rate larger than the economy assumes that the company will, one day, be larger than the economy. In practice, using such a low "g" value for fast growing companies will lead you to undervalue fast growing companies to the extent that you would never think that they are undervalued. In other words, you would be taking too large of a margin of safety.

Since no one really knows when it will slow down, it is best to assume that it will slow down sooner (after the 4th year) rather than later (many years into the future). This prevents us from ever overestimating the perpetual growth rate for fast-growing companies and prevents us from overpaying for a business. In other words, by assuming the growth will stop sooner, we are taking an **important** margin of safety for us because no one knows when they will stop growing so quickly.

    Before calculating the perpetual growth rate for Best Buy, we need to remove outliers. When looking at historical data, it is important to check for years where there is a dramatic change or an outlier that does not represent what will happen in the future. These are unlikely to reoccur in the future, so we do not want to consider them when estimating the firm's perpetual free cash flow growth rate.

    When we are looking at the free cash flow growth rates, 2017 sticks out to me. During 2017, they dramatically increased their free cash flow. The increase mainly occurred because their 2016 free cash significantly was lower. In 2016, they merely had $0.69 billion in free cash flow. Since such growth is very unlikely to reoccur, we would consider 2017's growth an outlier. In the future, it is unlikely that Best Buy will have an 184% growth rate in their free cash flow. Therefore, we are going to exclude it when we are taking the average of the free cash flow growth rates. Looking at the growth rates in the table, which I have recopied below, we do not see any other major outliers. Now, we are ready to start predicting the "g" value. But before we do this calculation, it is always good to ask yourself the following question: does this business have a moat?

	Historical Performance				Estimations			
Year Ending Feb	2017	2018	2019	2020	2021	2022	2023	2024
Free Cash Flow	1.963 Billion	1.453 Billion	1.589 Billion	1.822 Billion	1.438 Billion	1.468 Billion	1.497 Billion	1.526 Billion
Free Cash Flow Growth Rate	184%	-25.9%	9.3%	14.6%	-21.0%	2.08%	1.9%	1.9%

If you have not read the chapter about finding wonderful businesses, I would highly encourage that you do before you buy another stock. It is a must-read chapter. In that chapter, we stated that a wonderful business has a moat and that it is operating in a predictable industry. By investing in slowly changing industries and in businesses with moats, free cash flow will be predictable; it is unlikely that there would be any factors that could hurt a firm's free cash flow and make it significantly different from past years and significantly different from our expectations. These characteristics of a wonderful business ensure that the average historical and predicted growth rate will be similar to its perpetual growth rate; since the firm and industry are barely changing, their free cash flows would also barely change from year to year, allowing us to use historical averages to predict the future.

Once we have assured ourselves that we are investing in a slowly changing industry and in a business with a strong moat, and once we have gotten rid of outliers, we can start to calculate the perpetual free cash flow growth rate.[48]

We can calculate it by simply taking the average of the expected and historical free cash flow growth rates since it will be consistent over most years for predictable businesses that have a moat.

For Best Buy, the calculation would be the following:

$$"g" = \frac{-25.9\% + 9.3\% + 14.6\% - 21.0\% + 2.08\% + 1.9\% + 1.9\%}{7}$$

	Historical Performance				Estimations			
Year Ending Feb	2017	2018	2019	2020	2021	2022	2023	2024
Free Cash Flow	1.963 Billion	1.453 Billion	1.589 Billion	1.822 Billion	1.438 Billion	1.468 Billion	1.497 Billion	1.526 Billion
Free Cash Flow Growth Rate	184%	-25.9%	9.3%	14.6%	-21.0%	2.08%	1.9%	1.9%

The average of the growth rates is equal to -2.44%. We can use Best Buy's perpetual free cash flow growth rate or "g" value of -2.44% when calculating the terminal value: $\frac{FCF4(1+g)}{r-g} * \frac{1}{(1+r)^4}$

---

[48] I'll be honest. Best Buy does not have the strongest moat. Why should someone buy a product from Best Buy instead of somewhere else? They don't sell anything that is proprietary, and their brand is not the best in my opinion. They also do not have a good profit margin. But for the purposes of this book, it makes a good example because it has a recent outlier, which many well-known stocks do not have.

As we can see, Best Buy's perpetual free cash flow growth rate or "g" value is negative, which means that we can expect the company's cash-generating ability to start shrinking after 2024. This is generally a bad sign for a company. It shows that their ability to generate cash is shrinking. But it does make sense in this scenario. Retail has already peaked as they are losing market share to e-commerce. There is also currently a global COVID-19 pandemic affecting the business, which is contributing to a lack of growth. And, this pandemic may have long-term effects on the business. As of the time I am writing this book, no one knows whether there will be long term effects from the pandemic as it is still a developing situation. I try not to invest in companies when we can expect their cash flows to shrink over the long term because I prefer to invest in growth. When you are investing in companies where their free cash flow is growing, it is much easier to make money in the stock market.[49]

Anyways, looking back at our equation to calculate intrinsic value, we can see that we have calculated all of the variables to calculate the intrinsic value of a company. Now, we simply need to plug in numbers to calculate the intrinsic value of Best Buy.

Intrinsic value
$$= \frac{FCF1}{(1+r)^1} + \frac{FCF2}{(1+r)^2} + \frac{FCF3}{(1+r)^3} + \frac{FCF4}{(1+r)^4}$$
$$+ \frac{FCF4(1+g)}{r-g} * \frac{1}{(1+r)^4}$$

---

[49] Best Buy is also a good example for this book because it shows that companies don't always grow, as some beginners may assume. Some people have the idea that all stocks will go up in the long run. Best Buy is a good example because it shows that some companies are not in a position to do well over the long run.

$$\$12.86\ Billion = \frac{\$1.438}{(1+.1)^1} + \frac{\$1.468}{(1+.1)^2} + \frac{\$1.497}{(1+.1)^3} + \frac{\$1.526}{(1+1)^4}$$
$$+ \frac{1.526(1-.0244)}{.1+.0244} * \frac{1}{(1+.1)^4}$$

After plugging in the variables, we get Best Buy's intrinsic valuation or market capitalization to be about $12.86 billion. We would want to invest in this stock if it comes below that market capitalization.

We can also use the intrinsic market capitalization of a stock to calculate the intrinsic price per share of the stock. Although it is not necessary to calculate the intrinsic value of an individual share, it can be more beneficial to you because the share price is more often reported than market capitalization.

Remember that market capitalization is simply equal to the number of shares outstanding times the stock price. To figure out the price which we want to invest in a company, we can simply divide the intrinsic market capitalization by the number of shares outstanding. The number of shares can be found on the first few pages of an annual report. It will be found at the bottom of a page titled "United States Securities and Exchange Commission," and this page is usually located right before the table of contents. According to Best Buy's most recent annual report, they have 256,971,220 shares of stock. After dividing the intrinsic market capitalization by the number of shares outstanding, we get the perfect price to invest in Best Buy: about $50 per share.

Right now, Best Buy is trading at about $77 per share. So, it is way above its intrinsic value, and I would definitely not invest in it right now. After calculating a stock's intrinsic value, you may find that it is currently way overvalued. That is okay. You will notice stocks will be overvalued more often than they are undervalued. If stocks were usually undervalued, then everyone would get amazing returns by just guessing what and when to buy. The key to investing is to be patient and to wait for the times when the deals are excellent.

In fact, just a few weeks before the time I am writing this, Best Buy was trading just at about its intrinsic value, a low of $48.10 per share.

But let's say that the stock came to a low of $51 per share, $1 above its intrinsic value. If I was looking to invest in this company, that $1 difference would have not prevented me from initiating a position because it is such a small difference (assuming that I like their moat, that they have a better profit margin, and that their free cash flow is expected to grow). In some books, people are going to say that you need a 50%+ margin of safety before you invest in a stock.[50] That is simply ridiculous. If you are confident about a stock's future cash flows and if the business is a wonderful business, then there is no need for such a large margin of safety. If you did take such a large margin of safety, you would rarely find an investment opportunity because stocks don't really become that cheap very often. The only reason it makes sense to take a large margin of safety is when you are very unsure about the company's future cash-generating ability. But in that case, it is no longer a wonderful business because investing in an uncertain future and unpredictable cash flows is just like gambling. Here is what Warren Buffett has to say about margins of safety:

---

[50] A 50% margin of safety refers to investing in a stock when it is trading 50% below its intrinsic value. This is too large of a margin of safety and unnecessary for businesses with predictable futures, especially since we take reasonable, smaller margins of safety within our calculations by rounding down.

"We favor the business where we really think we know the answer, and therefore, if a business gets to the point where we think the industry in which it operates and the competitive position (i.e. moat) is so chancy that we can't really come up with a figure, we don't really try to compensate for that sort of thing by having some extra large margin of safety. We really want to go on to something we understand better. So if we buy something like See's Candy as a business or Coca-Cola as a stock, we don't think we need a huge margin of safety because we don't think we are going to be wrong about our assumptions in any material way… We'd love to find them when they're selling 40 cents on the dollar, but we will buy those at much closer to a dollar on the dollar."

In essence, huge margins of safeties are not necessary in the stock market, and simply using conservative estimates throughout the calculation for intrinsic value will suffice as a margin of safety.

I know that there are a lot of steps to calculating the intrinsic value of a stock, the second step of becoming a successful investor, so here are the steps listed in order. If you need a moment to review right now or when you are valuing stocks, these steps are a good summary for calculating the intrinsic value of a stock.

1. Create a table to organize your work, as previously shown in the book.
2. Estimate a stock's future revenue for the next 2 years by using publicly available revenue estimates on stock research websites like Yahoo Finance

3. Use historical and estimated revenue growth to predict a stock's revenue for the following 2 years
4. Use the stock's historical profit margin to estimate their future profit margin. Use their estimated profit margin and revenue estimates to predict their future net income
5. Calculate the historical free cash flow to net income ratio and take its average in order to estimate what it will be, on average, in the future
6. Use future net income estimations and the average free cash flow to net income ratio to estimate future free cash flows for the next 4 years
7. Determine an appropriate perpetual growth rate, following the rules outlined in this chapter
8. Plugin the variables and a 10% "r" value into the equation to calculate the wonderful price to buy a stock

$$\frac{FCF1}{(1+r)^1} + \frac{FCF2}{(1+r)^2} + \frac{FCF3}{(1+r)^3} + \frac{FCF4}{(1+r)^4} + \frac{FCF4(1+g)}{r-g} * \frac{1}{(1+r)^4}$$

9. Buy a stock if it is trading below or at its intrinsic value.

Here is one note on calculating the wonderful price of a stock. This is a short but important message to hear. Sometimes, it can feel impossible to find a good deal when the markets are overvalued. So, it can seem pointless to be calculating the intrinsic value of stocks at that time because you won't find a good deal easily. However, it is still beneficial to be valuing stocks when you have the chance, even if the markets are high. If you value a company that you like but find that it is currently overvalued, write down its intrinsic value in a spreadsheet. So, when stocks start to become undervalued, you know the exact price to invest in it. Investing opportunities are only open for a short period of time. When they are open you have to act swiftly (while still being rational). Calculating intrinsic values ahead of time will allow you to be ready for times of opportunity and will allow you to make decisions swiftly. You won't have to worry about valuing many companies quickly. Instead, all you would need to do is hit the buy button when the time is right.

Also, this book sets high standards for your stocks because I never want you to buy a bad stock. Investing in bad stocks is just throwing money away. Due to the high standards, it will take time to find wonderful businesses at a wonderful price.

The Truth

I just provided you with an equation to calculate the intrinsic value of a stock. Warren Buffett believes that estimating free future cash flows and calculating their net present value is how you determine the value of a stock. However, he never actually writes out the equation and applies a terminal value. Here is what he said during a 1995 shareholder meeting:

> We are trying to look at businesses in terms of what kind of cash can they produce if we are buying all of them or will they produce if we are buying part of them, and there is a difference. And then at what discount rate do we bring it back... despite the fact that we can define that in a very kind of simple and direct equation, you know, we never actually sat down and written out a set of numbers that relate that equation. We do in our heads in a way obviously. That's what it's all about. But there is no piece of paper and there never was a piece of paper... So it would be attaching a little more scientific quality to our analysis than there really is...

we really like the decision to be obvious enough that it doesn't require making a detailed calculation, and it's the framework, but it's not applied in the sense that we actually fill in all the variables."

He is essentially saying that he believes in the accuracy of the equation that I just taught you. If your inputs are good, then your valuation of the business will be accurate. However, he and I never physically do the math and calculate the terminal value, perpetual growth rate, and the free cash flow to net income ratio.

Instead, the calculation is done in our heads. We first look at the market capitalization or valuation of the company. Then we look at the cash flow statement to get a rough idea of how much free cash flow the company has been producing. *After asserting that this is a wonderful company,* we then think about how much free cash flow the company will **roughly** produce in future years, when will the company produce the free cash flow, and whether the future free cash flows will allow us to get the desired return on investment. We also think about the probability that we are correct about our estimates (investing in wonderful businesses makes this probability high). If we think that the company will **obviously** produce enough cash so that we can get the desired rate of return, then we invest in it.

When I say that "we do the calculation in our heads," I do not mean that we are mathematical geniuses and are able to do the entire process of calculating intrinsic value in our heads. Instead, I mean that we are manually estimating a company's future free cash flows using our best judgement and deciding whether those future free cash flows justify purchasing the stock at its current valuation and are durable against the competition. For example, if a *wonderful business* has been producing roughly $1,000,000 in free cash flow over the past few years, can be expected to generate at least $1,000,000 in future years, is expected to have solid growth in upcoming years, and currently has a market capitalization of roughly $10,000,000, the decision to invest in this company is pretty obvious. We would want to invest in it. Obviously, if the company is growing faster, manually estimating future free cash flows is going to be more important than just assuming that the company will be producing at least the same amount of free cash flow in the future. But you get the idea. We try to take a more holistic approach, and invest in companies where we think that we have a general idea of the future free cash flows without doing the math and when we think that the decision to invest is obvious enough that we don't need to run the numbers.

So, is the formula for intrinsic value incorrect? No. There is nothing wrong with it. It is just a more mathematical way of doing the calculation that Warren Buffett and I do in our heads.

However, I do think it is better for you to do the calculation in your head as you get more comfortable with investing in the stock market. Manually estimating free cash flow is much quicker and easier than going through the entire process. Also, it is my personal preference to invest in companies when the decision is obvious enough that it does not require to go through the entire mathematical process.

Step 2 is Determining a Wonderful Price: Contract

The second most important step when investing is to buy a business at a wonderful price. You need to buy businesses when the present value of all the future free cash flows are worth more than the valuation of the business. Doing this will allow you to make lots of money in the stock market. But, in order to make lots of money, you need to agree and understand the following:

I _____ (insert name) agree to adhere to the following principles of Warren Buffett in order to become successful at investing:

- Intrinsic value can be calculated by using the formula in the chapter
- It is beneficial to take slight margins of safety by rounding down throughout the calculation
- The intrinsic value of a stock can be estimated by forecasting future free cash flows and discounting them at 10%

Making Dividend Investing Powerful: Summary

People either love or ignore dividends. Some people like them because dividend investing creates passive income, while others are not attracted to dividends due to their low returns. Although dividend investing does not lead to the best returns using the strategies most people use, it is definitely a viable strategy. In this chapter, we will go over how to use discounted cash flow analysis to calculate the intrinsic value of a dividend stock. This will not only allow you to calculate the intrinsic value of a dividend stock, but it will also allow you to outperform the market through dividend investing. This chapter, whether you are a dividend investor or not, will make dividend investing powerful.

Topics in this chapter include:

- The basics of a dividend
- The mistakes most people make when dividend investing
- A formula to calculate the intrinsic value of a dividend stock
- What to look for in a dividend stock

Making Dividend Investing Powerful

"Be fearful when others are greedy. Be greedy when others are fearful." -Warren Buffett

The great part of investing is that you get to take control of your money. You get to choose how to invest it. If you desire passive income, you can try dividend investing. Another fundamentally acceptable way you can invest is by predicting future dividend payments and calculating their present value. It is a very easy and predictable strategy to get lots of gains and to calculate the intrinsic value of a stock. It is also very appealing to people who want a passive income portfolio.

Dividends are very special in the modern investing world. Many investing gurus love to build a dividend investing portfolio in order to build up a steady stream of passive income. It is a viable strategy in the stock market because it can allow you to make lots of money. However, the way many people go about dividend investing is absurd. They buy dividend stocks simply if the stock has a nice yield, low payout ratio, and good dividend history. They simply guess a good price to invest in the stock. But how can you be certain that you are going to get a good return without knowing what it is worth today? Without conducting a full and proper analysis, you will end up throwing a lot of money at bad dividend stocks. So, this chapter will tell you what to look for in a dividend stock and how to calculate the intrinsic value of a dividend stock. Even if you don't classify yourself as a dividend investor, it is still important to know how to identify a safe and strong dividend because many of your stocks will conveniently be paying dividends. Let's first go over some of the basics and criteria of a good dividend stock before we analyze how dividend investing can be overpowered.

The Basics of a Dividend

If you are new to investing, it would be good to start off with the basics of dividend investing. Some stocks are going to pay a dividend. It is literally a portion of the company's profits that are distributed to shareholders. The dividend is **usually** paid out on a quarterly basis. So, if a company pays $4 in dividends every year, you will get $1 every 3 months per share of stock that you own. Dividends are always optional to pay. A company can decide at any time to start paying a dividend or to stop it (however, most companies do have a good track record of consistently paying a dividend). Companies usually start paying dividends when they have an excess of cash and don't have many opportunities to invest cash to grow the business. But dividend-paying companies do face a huge downside. When they pay their dividend, they are taking cash right out of their business. So, they have less money to reinvest and grow the actual business.

    The metric that many investors care about most is the dividend yield. The dividend yield is calculated by dividing the dividends a share pays by the price per share ($\frac{Dividends\ per\ share}{price\ per\ share}$). The metric is represented as a percent, and it shows the annual return that you can get solely from the dividend, assuming that the company does not grow nor cut the dividend. For example, if a stock currently has a dividend yield of 2.5%, you will get a 2.5% return before considering capital appreciation (i.e. before you consider that their stock will go up or down). The dividend yield also represents how much dividends you will get if you invest $100 in the stock. For example, investing $100 in a stock with a 2.5% yield will pay you $2.50 in dividends each year, and a stock with a 4% yield will pay you $4 every year for every $100 that you invest in it. In short, a high yield means that you are getting more money in dividends for every dollar that you invest.

Many new investors only focus on the dividend yield because, in general, a high dividend yield leads to higher returns from a dividend. However, this is not a great idea because the current yield does not indicate whether your long-term returns from a dividend stock will be good. Your long-term return from a dividend is made up of 2 factors: the current dividend yield and how it will grow in the future. It is all about focusing on getting a good combination of dividend growth and yield. For example, if a company currently has a small dividend yield, but is growing its dividend rapidly, it can potentially provide higher returns than a high-yield dividend stock with little dividend growth. The inverse is also possible. A high yield stock that is growing its dividend at a slow pace might have a higher return than a stock that has a low dividend yield and that grows its dividend at a fast pace. Essentially, there is no perfect dividend yield or dividend growth rate. It is more about having a good combination of a starting yield and future dividend growth. We will look at how to determine whether a stock has a good balance between dividend growth and yield later in this chapter.

The Rule for Payout Ratios

A metric that can help you determine whether a dividend is going to grow in the future is the payout ratio. It is calculated by dividing the amount of money the firm pays in dividends by the company's annual earnings($\frac{Dividends\ Per\ Share}{Earnings\ Per\ Share}$). It represents the portion of company earnings that are paid out in dividends. For example, a firm with a payout ratio of 45% distributes 45% of their net income to shareholders.

Typically, as the payout ratio increases, it becomes harder for management to grow the company because there is less cash available to reinvest. Also, as it increases, it becomes harder for a company to grow the dividend because there is less money left to distribute. For those reasons, you never want any of your companies to be paying too much of their net income out as dividends. In other words, you need to avoid high payout ratios.

A good rule of thumb is to never invest in companies with payout ratios above 70%. The reason why you would want it to be below 70% is that you would want there to be room for the firm to grow the dividend. If the payout ratio is much higher than 70%, then the firm is unlikely to grow its dividend at a significant rate because they are already paying out a lot of their net income. On the other hand, if it is below 70%, the firm still has a lot of cash flow available to grow the dividend.

High payout ratios are just very risky for long-term dividends. They are very susceptible to being cut during hard times. If a business is slowing down, there are very few major costs a business can easily and quickly cut in order to free up cash flow. Cutting any costs in large businesses requires serious time and planning. However, dividends are easy to cut because they are optional to pay. In fact, dividend payments can represent over 50% of a firm's earnings, meaning that it sucks a lot of cash out of businesses. Due to these reasons, many firms will rationally resort to cutting their dividend if times become uncertain, especially if the payout ratio is high.

High payout ratios also hinder growth. Typically, as the payout ratio goes over 70%, the firm has very little cash to reinvest in the company in order to acquire new customers and grow. If a company cannot grow due to a lack of necessary cash flow, the value of the business (and hence the stock price) will not increase much over the long term. At the end of the day, it does not matter if you are getting a nice dividend if the stock price is not doing well. So, don't forsake the business for a good dividend yield. Always keep in mind that payout ratios above 70% are putting you in a vulnerable position.

Analyze What Pays for the Dividend

In order for a company to grow their dividend for many years, they also have to be able to grow their earnings. If earnings are not growing, the dividend growth would eventually need to slow down or stop.

You can use Zacks.com to ensure that a company is growing earnings at an appropriate rate. Zacks.com is a stock research site. If you type in the ticker symbol of a stock, there will be a metric called "expected EPS Growth (3-5yr)." The metric represents the average rate analysts think that the firm's earnings will grow per year over the next 3-5 years. For example, if the metric is 6%, it means that analysts think that the firm will grow earnings at a rate of 6% per year over the next 5 years.

From my experience, an expected growth rate of 4% per year is good to ensure that the dividend can consistently grow in the future. If I see a dividend-paying company that is not expected to grow their earnings by much at all, I would not take a chance on it. There is just too much risk. If I invested in such a stock, I would be stuck in an awkward position many years down the line. The company would be in a position where they can't grow their dividend without making the payout ratio too high. So, have a standard for your dividend companies and ensure that their earnings are expected to grow along with the dividend.

The Intrinsic Value of Dividends: When Dividend Investing is Overpowered

In the last chapter, we used discounted cash flow analysis to calculate the intrinsic value of a stock. In other words, we calculated the present value of the cash that the business will generate. Similarly, we can calculate the present value of the dividends that a business will pay.

    If you have been in the stock market for a while, you were probably aware of the criteria that I just went over for a safe dividend. The truth is that there is not too much to analyze when deciding when a dividend is safe because companies are so committed to growing and sustaining their dividend. However, just because a stock has a good payout ratio and expected earnings growth, it does not mean it is a good investment. Those are prerequisites for a good dividend stock, but just having those qualities will not make it a successful investment. You still have to follow the two-step strategy: buy a wonderful business and buy it a wonderful price. You still have to make sure that the business has a moat, is in a slow-changing industry, and is financially healthy. You also need to buy the business at a wonderful price. If I were to guess one factor that causes most dividend investors to not outperform the market, I would say that the factor is that they don't buy their stocks at a wonderful price. They usually guess and say that "it looks like a good investment right now." I don't want you to buy a stock because it *looks* like a good investment. You have to *know* that your stocks are a good investment. So, we are going to look at how you can calculate the perfect price to buy a dividend stock. If you have not read the last chapter, you should. It goes over the basics of calculating intrinsic value (i.e. the price we should pay for the cash that an asset will generate). We will be using a similar process to calculate the present value of the dividends that a stock will generate; the process will allow us to know which stocks will pay enough dividends in order to make them a good investment.

A big realization I had when I started investing is that if you have a rough idea of how much cash a business or any financial asset will produce in its lifetime, you can determine a good price to pay for the asset. Simply, understanding this realization and the following formula will put you ahead of 99% of stock market investors. Here is formula that is similar to the one in the last chapter and that will allow you to calculate the intrinsic value of a dividend stock:

Intrinsic value
$$= \frac{D1}{(1+r)^1} + \frac{D2}{(1+r)^2} + \frac{D3}{(1+r)^3} + \frac{D4}{(1+r)^4}$$
$$+ \frac{D4(1+g)}{r-g} * \frac{1}{(1+r)^4}$$

D1= Dividend Payment in Year 1
D2= Dividend Payment in Year 2
D3= Dividend Payment in Year 3
D4= Dividend Payment in Year 4
r= Rate of Return
g= Perpetual Dividend Growth Rate

$\frac{D1}{(1+r)^1}$ takes the dividend payment that the company will produce next year, and it discounts it so that you can get a desired return on investment. The same is true with subsequent terms in the equation. As in the previous chapter, $\frac{D4(1+g)}{r-g} * \frac{1}{(1+r)^4}$ discounts all dividend payments after the 4th year and calculates their present value, what they are worth today. So, if we know how much a stock will pay in dividends, then we can determine how much we should pay today for access to the future dividends, given our required rate of return. So, our challenge when determining the intrinsic value of a dividend stock is going to be predicting its future dividend payments. Let's walk through an example of predicting the future dividend payments of a stock so that you can replicate the steps when you are analyzing dividend stocks.

First, We Need to Predict Dividend Payments for the Next 4 Years

Let's go through the steps of valuing a dividend stock, using Johnson & Johnson as an example. The first step to calculate the intrinsic value of a dividend stock is to predict the stock's future dividend payments, specifically for the next 4 years.

Predicting free cash flow can be quite tricky because it only depends on a company's performance. Their performance can change due to market conditions (but a rough estimate is predictable when a company has a moat). However, estimating a firm's future dividend payments is really quite simple because a vast majority of stocks pay their dividends consistently and grow it predictably. The management's decisions with the dividend are very predictable and consistent as long as there are no major disruptions in the industry.

In order to predict a firm's future dividend payments, we can use the past history of the stock. Specifically, we can use a firm's historical dividend growth rate to predict how the stock will grow their dividends in the future. Let's take a look at how we can use dividend history to predict Johnson and Johnson's future dividend payments.

Johnson & Johnson, a stock that has been paying their dividend and growing it for over 57 consecutive years, currently pays a dividend of $4.04. Their stock price is currently at $150. Over the past 5 years, the company has been growing its dividend at about 6.3% per year. Over the last 10 years, the dividend has been growing by about 6.87% per year. Over the last 3 years and over the past year, the company grew its dividend by about 6%. Given that the company has been growing its dividend by about 6% per year, it is safe to assume that they will grow it by 6% per year in the future. When picking to analyze Johnson and Johnson for this book, it ended up being a coincidence that they have been growing the dividend by about 6% for the past 1 year, 3 years, 5 years, and decade. Most stocks will not be that way, but many will be pretty close. You are going to want to focus on how they have been growing their dividend over the past 3-5 years. The past 3-5 years gives a good representation of how the company has recently been growing its dividend, and how the management will grow it in the future.[51] If a company has been growing their dividend by X% per year over the past 3-5 years, they are likely going to grow their dividend by about that amount in the future. If you take a look at the dividend histories of many of the companies, you will be amazed at how consistently the management teams have been growing dividends by a similar amount each year. The consistent dividend growth makes it easy to estimate future dividend payments, especially when companies have moats and grow their earnings at a steady rate.

 Going back to Johnson and Johnson, we are able to know that the dividend growth rate will not slow down because they only have a payout ratio of 52%.[52] They have plenty of room to grow their dividend

---

[51] When researching dividend stocks, I personally like to use seekingalpha.com because they provide all the information that you need to know about a dividend in a nice layout.

[52] Most companies can grow their dividends easily with a 52% payout ratio. From my experience of analyzing stocks, a normal or average payout ratio that you can expect from companies is about 60%. When payout ratios get above

and still have excess cash; since their payout ratio is not high, they will be able to continue to grow their dividend at the rate that they have been growing their dividend. Thus, it is realistic that the company will continue to grow their dividend at 6% per year.

Knowing that the firm will be able to grow their $4.04 dividend by 6% per year, we can predict their dividend payments for the next 4 years. So, to estimate the firm's dividend payments for next year, we simply need to multiply $4.04 by 1.06, the 6% growth rate. Then, to calculate the dividend payment for the second year, we simply need to multiply 1.06 by the dividend payment from the previous year. Similarly, to calculate the dividend payment in the third year, we simply need to multiply 1.06 by the dividend payment from the second year. I have entered these values in the table that is below.

Year	Year 1	Year 2	Year 3	Year 4
Annual Dividend Payment	$4.28	$4.53	$4.81	$5.10

The predicted dividend payments over the next 4 years are $4.28, $4.53, $4.81, $5.10.[53] In order to calculate the intrinsic value of a

---

70%, you can start to notice a considerable slow down in the dividend growth rate. That is one reason why you want to invest in stocks with payout ratios under 70%.

[53] One of the characteristics of a wonderful business is that it has predictable free cash flows. The same rule applies for a dividend stock. A wonderful dividend stock has a predictable dividend. If you ever feel that a company's future prospects and dividend are so uncertain that you do not feel

dividend stock, we need to predict all of the firm's future dividend payments and not just the first 4 years.

Next, We Need to Apply a Reasonable, Perpetual Growth Rate

In order to predict a firm's future dividend payments over the span of its lifetime, we can apply a conservative perpetual dividend growth rate to the dividend after the 4th year. This will give us a good estimate of the firm's future dividend payments over the long run.

So, what is the ideal perpetual growth rate for a dividend stock? This metric is relatively easy to figure out. We can simply look at how many stocks have been growing their dividend over the long run in order to predict how they will grow their dividend in the future. Over the past 50 years, stocks in the S&P 500 have been growing their dividend by about 5% per year.[54] For most stocks, this would be a good perpetual growth rate. However, since we want to take a margin of safety in our calculations, a safe, conservative perpetual growth rate would be 4% for most stocks. For most stock, this is a very good perpetual growth rate because a vast majority of companies can grow their dividend by at least 4% per year in perpetuity, as long as they have the characteristics of a good dividend stock and a wonderful business.

The only time when we would not use 4% as the perpetual growth rate is when the stock's average, historical, dividend growth rate

---

comfortable with your dividend estimation, you should move onto investing in a company that is more predictable.

[54] This metric was provided by Seeking Alpha. The article used data provided by NYU's Stern School of Business.

over the past 3-5 years is between 3%-4%. For example, if a stock has been growing their dividend by only 3%-4% per year over the last 3-5 years, it would be better to use a perpetual growth rate of 1%-2%. We do not want to assume that the company will all of a sudden decide to increase their dividend at a faster rate. We want to be safe with our hard-earned cash, so we do not want to overestimate the firm's future dividend payments by using an inflated perpetual growth rate. Using an inflated perpetual growth rate would cause us to overvalue a stock, preventing you from getting a good return. Also, if a stock has been growing its dividend at a rate less than 3% per year, I really would not want to invest in it because its dividend would not have much growth potential.

    But you are probably wondering why we can assume that the company will continue to grow their dividend by 4% per year. How do we know that the dividend won't be cut all of a sudden? How do we know that the dividend will continue to grow and not stay flat? These are all very valid points of concern. In all honesty, there is no way to be 100% certain that the dividend will continue to grow at that rate. An unpredictable event (i.e. a global pandemic) could occur and completely change the situation. No one has a crystal ball that tells us the future. However, we do have methods in order to ensure that a company will be able to easily grow their dividend.

    As explained earlier, it is crucial to invest in businesses that are expected to grow their earnings by at least 4% per year. A stock that has a 4% earnings growth rate will always have the potential to grow their dividend by at least 4%. You must also invest in companies which have payout ratios below 70% so that it is easy for the company to grow their dividend.

    So, by ensuring that a company does not have a high payout ratio and is going to grow its earnings at a good rate in the future, you are ensuring that the chances of a dividend cut are extremely low and that a healthy dividend growth rate is feasible. Ensuring that a stock is unlikely to cut their dividend or stop growing it is very important. When you apply a perpetual growth rate, you are assuming that the stock will continue to grow their dividends far into the future at the perpetual growth rate. If the company cuts their dividend or grows it by less than expected, your perpetual growth rate estimate would not be reflective of the company's future dividend payments. Then your calculated intrinsic value would be higher than what the company's dividends are actually worth. This would make you overpay for a stock, and you would end up with a lower rate of return than expected.

Once we have ensured that the stock is expected to grow their earnings by at least 4% and that it does not have a high payout ratio, we can apply an appropriate growth rate. Going back to the example of Johnson and Johnson, the appropriate perpetual growth rate would be 4%. They have been growing their dividend by 6% per year, which makes it safe to assume that their long term growth rate would be at about 4% because the dividend growth rate tends to fall at about 4%-5% in the long run. Also, their payout ratio is at 59%, which is well below 70%. Their low enough payout ratio makes me comfortable to assume that Johnson and Johnson will be able to continue to grow its dividend by 4% per year over the long run.

Lastly, We Need to Input the Required Rate of Return

Now, we are getting close to calculating the intrinsic value of Johnson and Johnson. Taking another look at our formula for intrinsic value and the data table, we have almost all of the inputs in order to calculate the intrinsic value. We have predicted future dividend payments for

Johnson and Johnson, and we have determined that 4% is an appropriate perpetual growth rate for Johnson and Johnson.

$$\begin{aligned}Intrinsic\ value \\ = \frac{D1}{(1+r)^1} + \frac{D2}{(1+r)^2} + \frac{D3}{(1+r)^3} + \frac{D4}{(1+r)^4} \\ + \frac{D4(1+g)}{r-g} * \frac{1}{(1+r)^4}\end{aligned}$$

Year	Year 1	Year 2	Year 3	Year 4
Annual Dividend Payment	$4.28	$4.53	$4.81	$5.10

~~D1= Dividend Payment in Year 1~~
~~D2= Dividend Payment in Year 2~~
~~D3= Dividend Payment in Year 3~~
~~D4= Dividend Payment in Year 4~~
r= Rate of Return
~~g= Perpetual Dividend Growth Rate~~

The only other input that we have not determined is "r," the required rate of return. When discounting dividends, "r" is the rate of return that you want solely from the dividends.

What is the ideal required rate of return? I recommend using a required rate of return of 6% when discounting a stock's future dividend payments. Using a 6% required rate of return for discounting the dividends ensures that you will have a good long-term return from the dividends. If you get a 4% return from stock price appreciation per year (which is **very realistic** for companies that meet the criteria in this book), along with a 6% return from the dividend, you will predictably get a good return on investment.

To some people, a 6% return from dividends sounds very high because it is hard to find a good dividend stock with a 6% dividend yield. But when you go through the math, it is not very hard to find companies trading at their intrinsic value with an "r" value of 6% because dividends grow over time. Your money will produce more money each year, so in practice, it is very feasible to find good dividend stocks at their intrinsic value using a 6% "r" value, especially during times of panic.

Intrinsic value
$$= \frac{D1}{(1+r)^1} + \frac{D2}{(1+r)^2} + \frac{D3}{(1+r)^3} + \frac{D4}{(1+r)^4}$$
$$+ \frac{D4(1+g)}{r-g} * \frac{1}{(1+r)^4}$$

"g"=4%

Year	Year 1	Year 2	Year 3	Year 4
Annual Dividend Payment	$4.28	$4.53	$4.81	$5.10

Now, we can plug in all the variables and calculate the intrinsic value of Johnson and Johnson. After plugging and chugging, we get an intrinsic value of $226.21. Given that Johnson and Johnson is only trading at $150 right now, it looks like a good long-term dividend investment (i.e. it is undervalued).

Just to show that this formula calculates the present value of all the future dividend payments of a stock, I went on a spreadsheet and forecasted all the future dividend payments of this stock using the 4% perpetual growth rate and the 6% growth rate. I then calculated the present value for each year's dividend. I then took the sum of all the present values to get the same intrinsic value as when we used the equation. This just shows that the expression with the "g" value can be used to calculate the intrinsic value of a stock. For many people (including myself), the formula with "g" is not intuitive, while the spreadsheet is. So, I like to explain that individually discounting dividends and taking their sum is equal to the formula with "g" so that you can feel assured that the formula does in fact represent the intrinsic value of a stock.

But just because a stock is trading at or below its intrinsic value doesn't mean that you should buy it. Let's explore why in the next chapter about diversification.

Making Dividend Investing Powerful: Contract

Dividend investing is a viable investing strategy because your returns are very predictable. Where people go wrong with dividend investing is that they don't go through the process of calculating the intrinsic value of the dividends when making their decision. So, it is important to agree to the following:

I _____ (insert name) agree to adhere to the following guidelines of dividend investing:

- Wonderful dividend stocks have payout ratios under 70%
- A solid future earnings growth rate of 4% is helpful because it facilitates dividend growth
- In order to determine the present value of a dividend stock, we can forecast future dividend payments and calculate the net present value of the dividend payments
- I will actually go through the calculation in order to make a thought-out decision

When Diversification Hurts: Summary

Diversification, the process of spreading your risk across multiple investments, is highly praised in the investing world for an obvious reason; it protects your hard-earned money! However, diversification can also be dangerous for your portfolio. Whenever you buy a new stock, you are potentially allocating less money to a better stock. So, this chapter determines a balance between spreading your risk and concentrating your money in good stocks.

Topics in this chapter include:

- How diversification can kill your returns
- When it does not make sense to diversify
- Why Warren Buffett is not always in favor of diversification
- When it does make sense to diversify more
- Whether you should be diversifying more or less
- How index funds can be beneficial to your portfolio

When Diversification Hurts

"Diversification, as practiced generally, makes very little sense for anyone that knows what they are doing." -Warren Buffett

Diversification is a key aspect of any investor's success. However, there are good ways to diversify, and there are bad ways to diversify. This chapter will go over good strategies to diversify your portfolio and mistakes that many investors make when diversifying their portfolio.

The Basics of Diversification

For those who don't know, diversification is when an investor purchases many stocks that are mostly independent of each other and are significantly different from each other in order to mitigate risk; if one company or industry fails, you always have a backup.[55] It is avoiding having too much of your money in one stock, asset class, or industry because you don't want one company or one industry trend to kill your performance. Essentially, diversification prevents having all your eggs in one basket. There are two main ways diversification protects you:

1. Diversification mitigates your risk when you pick a bad stock.

---

[55] When I used the word "independent," I am referring to the idea that you don't want the success of two or more of your stocks to be reliant on each other or on similar factors. If one company starts to struggle, you don't want the other to start struggling at the same time.

2. Diversification protects your portfolio from adverse industry trends.

The first way your portfolio is protected by diversification is self-explanatory. When you diversify and own many companies, each investment or stock makes up a small portion of your portfolio. If you make an error in your analysis and purchase a bad stock, it would not cause much harm to your returns when it makes up a small portion of your portfolio. If it made up a large portion of your portfolio, then it would severely hurt your performance.

Diversification also protects you from adverse, industry-wide trends. When you diversify your portfolio, your money will be invested in many industries. Over time, the industries that you are invested in will change, and sometimes, these changes will hurt the stocks that you own. In order to prevent a portfolio from taking a big hit, you can own stocks in many different industries; if you are too heavily invested in one industry, and if that industry's growth starts to slow down, almost all of your stocks will start to have trouble growing the amount of cash that they generate. However, if all of your stocks are in different industries, one adverse trend in a single industry would only affect one or a few of your stocks. Thus, diversification reduces risk because it mitigates repercussions from adverse, industry-wide trends.

A Big Diversification Mistake

Many investing books and websites urge investors to not have too much of their portfolio in one sector. However, diversifying into multiple sectors is not important and can actually hurt your returns. In order to understand the proper way to diversify, you need to understand the difference between a sector and an industry. A sector is a portion of the economy. There are 11 sectors in the economy: financials, utilities, consumer discretionary, consumer staples, energy, healthcare, industrials, technology, telecom, materials, and real estate. An industry is a portion of a sector. For example, under the financial sector of the economy, there is the banking, insurance, and brokerage industries. *Sectors describe the part of the economy a business operates in, while industries describe a firm's operations within a sector.* Many investing books and websites will give "sector weighting" recommendations, which describe the portion of one's portfolio that should be invested in each sector. In fact, some people try to match the S&P 500's sector weightings, and I have seen others who try to have their money distributed equally across the sectors of the economy.

      However, you can have a well-diversified portfolio by investing in stocks that are in the same sector because the performance of stocks within the same sector are not necessarily related to each other. For example, in the consumer staples sector, you can purchase Proctor and Gamble and Pepsi shares. Proctor and Gamble is a company that sells common household products like Tide pods, Crest toothpaste, and Olay skincare products. Pepsi, on the other hand, is a completely different business. They mainly sell beverages and snacks. Although the two firms are categorized as being in the consumer staples sector, the two businesses are very different. The performance of Pepsi is not influenced by Proctor and Gamble and vice versa. Purchasing both of them will mitigate a portfolio's risk because consumer trends in the beverage industry do not affect Proctor and Gamble's products; if soft drinks and chips start to get unpopular, Proctor and Gamble's business would not slow down. Since they operate in totally different industries, you are protected from adverse industry-wide trends. Therefore, you should focus on diversifying across industries, rather than focusing on diversifying across sectors.

When You Should Not Diversify

Investing is simply the process of allocating money in order to safely and predictably achieve the highest return possible. Good investors are able to determine and identify the stocks that can provide them with the highest return without excessive risk.

Investing is all about opportunity cost. Let's say that I decided to spend a few hours of my weekend valuing stocks and calculating their intrinsic value so that I could invest $10,000. Also, let's assume that I found three stocks to be trading at or below their intrinsic value using an "r" value of 10%.

Company	Stock A	Stock B	Stock C
Current Share Price	$95	$75	$60
Intrinsic Value	$100	$125	$60

The intrinsic values and share prices of the three stocks are represented in the table. Stock A, B, and C are trading at or below their intrinsic value. If I am confident in my calculations and in the underlying business, how should I allocate my $10,000?

A person might say that it would be smart to invest in all three stocks and allocate roughly $3,333.33 to each stock. However, that does not consider **opportunity cost.** If I invest in Stock C, which is trading at its intrinsic value, the money that I allocated to Stock C would not get the higher returns that I could get by investing in stock B. Stock B would get significantly higher returns than stock C because stock B is trading significantly below its intrinsic value, while stock C is trading at its intrinsic value.

Stock C is worth about $60 per share, and we can get them for $60 per share. While, with stock B, we can pay $75 for something that we think is worth about $125. Doesn't stock B sound like a much better investment? So, it does not make sense to invest in stock C when I could invest in stock B. The same logic would explain why I would not want to invest in stock A. Stock B is trading at a higher discount. Why would I want to invest in Stock A or C when I know that I can invest in stock B and predictably get higher returns?

You want to purchase stocks that are trading the most below their intrinsic value. This phrase is key to understanding investing. The process of investing is allocating money in the best places available to us that do not impose excessive risk on the us. If we find an investment that is trading 10% below its intrinsic value and another stock trading 20% below its intrinsic value, we would want to invest in the stock that is trading 20% below its intrinsic value because it will lead to higher long term returns. In other words, the more a stock is trading below its intrinsic value, the better the deal because it is more undervalued.

Intuitively, it makes sense to maximize a portfolio's diversification because it would minimize risk. However, diversification comes at a cost. When you diversify your portfolio, you are adding less-amazing stocks to your portfolio. For example, if you create a list of the top 10 stocks to buy, and only decide to buy your top 5 stocks, you would be investing in companies that you think are the most attractive. These companies should ideally perform well because these are your top choices.

However, if you were to purchase all of your stocks on your top 10 list, stocks 6 through 10 would not perform as well as stocks 1 through 5. By allocating money to stocks 6 through 10, you would be lowering your returns because stocks 6 through 10 would underperform stocks 1 through 5. Investing in stocks 6 through 10 adds diversification, but it comes at a huge cost. You are forging businesses that will provide higher returns for businesses that will generate lower returns. Take a look at what Warren Buffett has to say about diversification:

> "Diversification, as practiced generally, makes very little sense for anyone that knows what they are doing. Diversification is a protection against ignorance. I mean if you want to make sure that nothing bad happens to you relative to the market you [should] own everything. There is nothing wrong with that. I mean that is a perfectly sound approach for somebody who does not feel that they know how to analyze businesses. If you know how to analyze businesses and value businesses, it's crazy to own 50 stocks, or 40 stocks, or 30 stocks probably because there aren't that many wonderful businesses and that are understandable to a single human being in all likelihood. And to have some super-wonderful business and then to put money in number 30 or 35 on your list of attractiveness and forgo putting more money into number 1 just strikes Charlie and me as madness."

There are many "financial experts" and gurus who don't properly diversify and destroy their portfolio's potential. For example, there are many dividend investors who own 50+ stocks and still aim to continue to invest in more businesses in order to diversify their portfolio. However, as those investors continue to invest in more stocks, their portfolio tends to have so many companies that the gain from the excellent companies, the less substantial gain from the decent companies, and the loss from the poor stocks cause the portfolio to perform like an index fund; when you invest in too many companies, you are essentially creating your own index fund. If you want to own 50 different companies, then you might as well invest in a low-cost index fund instead of spending valuable time researching so many companies and putting stress on yourself. With any investing strategy, it is important to make it your goal to beat the market because it will make the time spent researching worth it.

In order to properly diversify, you need to find a proper balance between mitigating risk and maximizing returns. If you do not diversify enough, you will constantly be worried about your investments and may potentially lose lots of money. If you over diversify, your returns will be compromised, and you will be unnecessarily forming an index fund.

A very common question investors have when building their portfolio is "how many stocks should I buy?" People expect me to be able to provide a magic number that applies to everyone. That simply does not exist. How many stocks would Warren Buffett own? Well, on that topic, he says the following:

> "Within Berkshire, I could pick out 3 of our businesses, and I would be very happy if they were our only businesses we owned, and I had all my money in Berkshire."

Warren Buffett would be perfectly happy, calm, and confident only owning 3 stocks. I would not simply copy him and only own 3 stocks. Warren Buffett has significantly more experience than you and I. Warren Buffett is a smarter investor than you and I. Warren Buffett is more emotionally prepared to handle volatility than you and I. Since he is a much smarter investor than both of us, he really does not need to diversify as much. "Diversification is a protection against ignorance," and we are more ignorant than Warren Buffett. So, you and I need some more diversification, especially if you are just getting started with investing.

Also, given your current financial situation, you might be forced to diversify. Most people have a job and receive money in increments. Let's say that you are able to save $1,000 every month for investing purposes. Since you are receiving your money in increments, you are going to naturally end up investing in many companies. For example, if you are only able to find one wonderful business at a wonderful price every few months, you would end up investing $12,000 in a few different stocks every year. Next year, you would have an additional $12,000 available to invest. Those same businesses are probably not going to be at a wonderful price next year, so you are going to need to invest in different companies. So, if your investment account is being funded over time, your portfolio will naturally become highly diversified.

On the other hand, Warren Buffett has all the cash that he could possibly need. As soon as he finds a wonderful business at a wonderful price, he can invest a meaningful amount of money because he has so much cash at his disposal. In other words, he does not need to make his investments in increments. Hence, you should not simply copy Warren Buffett's ideal diversification principle because his situation is very different from your situation.

An Ideal Number of Stocks

Although there is not an ideal number, here is a range that has a good balance between diversification and concentration in strong companies. I would recommend and be comfortable with a portfolio of about 5-15 different companies. For me personally, 5 companies is the bare minimum amount of stocks that I want to be invested in so that I can feel assured that my risk is spread across the success of multiple companies. Whereas owning 15 companies provides enough diversification for me. If your portfolio is starting to have more than 15 different stocks, I would recommend trying to be more picky in your choices and concentrating your money into the best investment opportunities that you can find.

Also, you should try not to have a single company represent more than 20% of your portfolio. When a single stock is too heavily weighted in a portfolio, your money would be too reliant on one investment. Based on my experience, 20% is a number that I am comfortable with. I feel confident in my ability to choose stocks; I don't believe that my money needs to be more diversified when I can just invest in companies that I can confidently say are trading below their intrinsic value.

But that 20% rule might not be the best for you. If you are new to investing and are only dipping your toes in the field, it would be smart to make your first investments be fairly small. With each stock representing a small portion of your total portfolio, you would be less reliant on a single stock. Starting off safe will prevent you from facing significant repercussions if you make a beginner's mistake or an emotional decision.

However, those are only **my** preferences. Your amount of diversification should **only** depend on your comfort level with the stock market. If you have been doing well and are confident in your decisions, feel free to diversify a little less. On the other hand, diversify a little more if you are constantly feeling worried and checking your portfolio (I believe that most people should be diversifying more than I diversify). If you feel very concerned when you pick individual stocks, you can invest

most of your money in an index fund. You'll get a return that will allow you to build large amounts of wealth. Then, you can invest a small amount of money in a miniature portfolio of individual stocks so that there is very little personal risk if you ever pick a bad stock.

That's the great part about investing; you have control over your portfolio. If you don't feel comfortable in your investing abilities, you can always diversify more in order to reduce the amount of risk. If you receive cash incrementally, then feel free to buy stocks more frequently than an investor like Warren Buffett who rarely buys stocks. Investing according to your needs will make you feel free because you are not following rules which are not the best for you.

Taking a Diversified Route

Although diversification can reduce your potential returns, it is beneficial for most people to over-diversify. Most people would have anxiety every day if their entire retirement was based upon their own stock picks. It is not a bad feeling. In fact, it is normal. However, if you feel comfortable with having most of your money in 10 or 20 stocks, you are in the minority that has the potential to get more returns than the majority. But everyone has a different risk tolerance, and I would imagine that most people are going to want to extremely diversify their portfolio. Once again, there is nothing wrong with that. It is possible to get rich through both methods. Although diversifying more will make your path to wealth slower, it is still very achievable for pretty much everyone. So, just know your risk tolerance.

If you are going to focus on maximizing diversification, here is how I would go about it in a way that still allows you to pick stocks and have lots of safety. Allocate 90% of the money that you are going to invest in an S&P 500 index fund. The S&P 500 is a collection of the 500 largest, American, publicly traded companies. Companies that are larger represent a larger portion of the S&P 500. When you are investing in an S&P 500 index fund, you are investing in those 500 companies, with more of your money being allocated to the larger companies. There is essentially no long-term risk investing in the S&P 500 due to the high amount of diversification, and you can get a good enough return so that you can achieve your financial goals.

Also, an S&P 500 index fund is a much better option than mutual funds, which most people own. If you don't know, a mutual fund is like investing in an S&P 500 index fund except that someone else chooses which stocks your money will be allocated to. With an S&P 500 index fund, your investment is passively managed, meaning that no one is choosing which stocks to buy and sell. Your money is always invested in the 500 largest, American, publicly traded companies. On the other hand, people who work at mutual funds analyze stocks and make investment decisions based on their own criteria. They can buy and sell stocks whenever. Intuitively, it sounds smart to invest in a mutual fund because you would think that they would get a much higher return than an S&P 500 index fund. However, mutual funds typically don't outperform the market.[56] They have to diversify to the extreme in order to ensure that their customers' money is safe. With so much diversification, they typically end up getting a return that is similar to or lower than the S&P 500. In most cases, mutual funds actually end up underperforming the market.

Not to mention that mutual funds charge higher fees than an S&P 500 index fund. For example, all mutual funds charge what is known as an expense ratio. Essentially, each year they will take the value of your investment and multiply it by the expense ratio. This

---

[56] Many funds get a return that is slightly below the S&P 500.

amount will represent an expense coming out of your investment. It covers costs associated with running the fund. So, if you have $100,000 in a mutual fund with an expense ratio of 1%, you will be paying $1,000 in fees to the fund due to the expense ratio. An average mutual fund can have an expense ratio around .7%. Although .7% sounds like a very small number, it is large, especially when compared to index funds. Many index funds have an expense ratio under .1%. In fact, Vanguard's S&P 500 ETF (ticker symbol: VOO) has an expense ratio of .03%.

The difference between .03% and .7% will make a huge difference in your long-term returns. If you invest $100,000 in an S&P 500 index fund that gets a 10% return per year (or 9.97% after the expense ratio) over 50 years, you would have about $11.5 million. On the other hand, if you invest the same $100,000 in a mutual fund that averages 10% per year (or about 9.3% after the expense ratio) over 50 years, you would have about $8.5 million.[57] That is a percent difference of 30% between the two final values.[58] Essentially, a small reduction in return per year over a long period of time will cost you a lot of money. Since index funds have much lower fees and usually get higher returns, they are much better long-term investments than actively managed mutual funds.

---

[57] The figures were calculated only taking into account the expense ratio. There are other fees associated with mutual funds. For example, funds will charge a fee when they make a trade. Certain funds also have a "load" which can be costly. I would never invest in a fund with a load. These calculations do not take inflation nor taxes into account. It also does not take into account that many mutual funds get lower returns than index funds. Since actively managed mutual funds have higher fees and usually get lower returns, an index fund is a smarter option in order to diversify your money.

[58] A percent difference between two numbers can be calculated by taking the absolute value of the difference between two numbers and dividing it by the average of the sum of the two numbers.

I'm assuming that you are like most people in America, that you have a job, and that you receive a paycheck every other week or so. This means that you have to steadily invest your money over time. If you are in such a situation and plan to invest in an index fund, I would recommend that you practice a strategy called dollar-cost averaging. It essentially means that you invest your money over time. You diversify your money by spreading the risk over time. After every paycheck, you can invest a part of your paycheck into an index fund. Although some people recommend investing a constant amount each time you get a paycheck, I recommend something else. If the markets are high, invest only a small amount of money into an index fund. When the markets start falling, invest more money into an index fund. Doing this will allow you to get higher returns; you will have a lower cost basis if you buy heavily when the markets are down.[59] It is a more efficient way of investing than just investing a constant amount of money whenever you get your paycheck. Also, most personal finance experts recommend investing 10%-15% of your gross income. However, it is only a general rule of thumb. The amount you need to invest will always depend on how old you are today, what your retirement goals are, and how much you have already invested. Nonetheless, it is important to consistently invest a meaningful amount of money. And once again, you want to invest 90% of your investment in an S&P 500 index fund if you want to maximize diversification and still pick stocks.

With the remaining 10% of the money, you can buy individual stocks. By creating a miniature stock portfolio with 10% of your money, you will be able to get some experience investing and the ability to create a miniature portfolio that has the potential to get high returns.

---

[59] The word "cost basis" refers to the average cost you purchased your shares at.

At the same time, your personal wealth will never be at much risk because most of your wealth will be tied to a diversified index fund that will get decent returns. Of course, you can tweak the ratio of money invested in stocks and the index fund as you wish. It all revolves around you. But, having 90% in an index fund and 10% in stocks is a good balance for investing in individual stocks and having very little anxiety.

    Here is the last thing to know about taking the diversification route. I was recently listening to Dave Ramsey's podcast on YouTube. He is a very smart personal finance expert, and I would highly recommend that you check him out. He was talking about a study which looked at the biggest factor that affects how much money people have when they retire. Before reading the study, he thought the factor would either be the returns or expenses of their mutual funds because a small change in these has a huge effect on returns. In reality, the study found that the amount of money you actually invest determines how much money you will have later. It seems pretty obvious, but many people are oblivious to it. If you invest more money today, you will have more money later. Investors are always too busy looking for the best place to put their money. When they find the best place to put their money, they don't actually put enough money into it in order to make a lot of money from it. When you invest in an index fund, you have to put a lot of money into it in order to make a lot of money. The more money you put in sooner, the more money you will have later. It is a very simple lesson, but it is very important to be conscious about it.

When Diversification Hurts: Contract

At the end of the day, your level of diversification depends on your risk tolerance and investing confidence. You have to make the right decision when diversifying your portfolio, whether it be to diversify a lot through an index fund or to diversify a little by picking the best stocks. But at the end of the day, the decision revolves around your comfort level.

I _____ (insert name) agree to following about diversifying a portfolio:

- *I have to diversify so that I am stress-free. If I am stressed out by the stock market, then I should diversify more or invest mainly in an index fund. If I feel comfortable with less diversification, then I will concentrate my money in the best stocks*
- *I will focus only on myself when diversifying. Even though it is easier to get higher returns through less diversification, I will not concentrate my money into a fewer amount of stocks unless I feel comfortable with the risk*

Emotions Kill Your Returns: Summary

Your emotions are your greatest enemy. Even when you know the right decision to make, they prevent you from making rational decisions. This chapter will go over how your mind can cause you to make bad decisions, even if you are a long-term investor. We will also go over how paying less attention to the markets can help you become a less emotional investor.

Topics in this chapter include:

- How a marketing technique called "anchor pricing" causes you to think stocks are undervalued when they are not
- Why you should never forget about the Monte Carlo Fallacy
- How trying to keep up with the Joneses is not always the best idea
- A method to reduce emotional decision making

Emotions Kill Your Returns

"The most important quality for an investor is temperament, not intellect." -Warren Buffett

Your emotions are your greatest enemy when investing in stocks. You need to understand how your emotions go against you when you are investing. If you can identify when your emotions are preventing you from making rational investment decisions, then you will be able to realize when you are leading yourself into making a bad decision.

Anchor Pricing

Anchor pricing is a very powerful marketing tactic. I can guarantee that this tactic has been used against you, as a consumer and as an investor. As the name suggests, it involves establishing an "anchor." In marketing, the anchor is a highly priced item that is placed next to a firm's main product. The anchor makes adjacent products seem cheaper than they really are. If you ever go shopping for a television at an electronics store like Best Buy, you are going to notice that they have a bunch of televisions. You are going to find some televisions that cost several thousand dollars, and you are going to see some televisions which only cost a few hundred dollars. In general, people don't purchase the televisions which cost several thousand dollars. So why do stores keep them?

Let's say that you are looking to buy a television that roughly costs $500. You walk into a television store, and you see that all of their televisions cost $400 or $600. As you are shopping, you decide to compare the benefits of televisions which cost $400 and $600. You notice that the quality difference between the two televisions is very minimal, so you end up just purchasing the cheaper $400 television.

Now imagine that the same store has televisions which cost $400, $600, and $1200. The quality differences between the $600 and $1,200 television are minimal. However, the $600 television would all of a sudden look fairly cheap due to the $1,200 television being present; the $1,200 television is acting as an anchor to influence your judgment on the $600 television. You would notice that the $1,200 and $600 televisions are very similar, but the $600 is significantly cheaper. This would make the $600 television look like an attractive purchase, even though having a $1,200 television next to a similar, $600 television does not make the $600 television a better purchase.

When you are investing in the stock market, the same principles of anchor pricing lead investors to make bad decisions. If you see a stock price start to fall, plenty of investors automatically assume that the stock is cheap. Why does the stock appear to be cheap? Well, some investors would say that the stock is cheap because it "looks cheap." It "looks cheap" because its price has come down significantly. The stock "looks cheap" because you are comparing the current stock price relative to what the same stock was trading at in the past; the higher, historical stock price is acting as an "anchor" for you to make a judgment on the current stock price.

Just because a stock or anything comes down in price or is cheaper relative to an alternative does not make it actually a good deal. Just because a $1,200 television (the anchor) is next to a similar, $600 television does not automatically make the $600 television any better of a purchase had the $1,200 television not been there. Similarly, if you go to the grocery store and see that bananas cost 40 cents per pound at regular price, you might be inclined to buy because they would seem like a good deal. On the other hand, if you saw bananas and that they were 50% off, but with the same final price of 40 cents per pound, you would be even more inclined to purchase them. The 80 cents per pound price (the regular price of the bananas) is acting as an anchor to make the discounted bananas look cheaper at 40 cents, even though you're getting the same object for 40 cents per pound.

You don't want to automatically think that a stock is cheap because its price is down. Just because a stock goes down, it does not mean that it will go back up. Just remember to focus on the cash that the business will generate relative to its price (market capitalization). When you see that a stock is cheap relative to historical prices, it is not any more of a good deal than if the stock had risen by 100% to come to the same price; in both scenarios, you would end up with the same asset, paid the same price for the same asset, and the same asset would have generated the same amount of cash in future years. So, it is just illogical to rely on historical prices to base investment decisions.

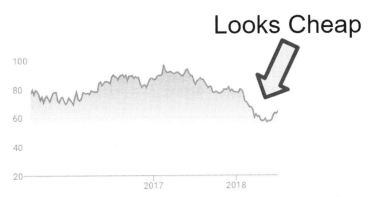

In this chart, the stock "looks cheap" because its price is low. People who rely on charts to make investment decisions would buy this stock right now.

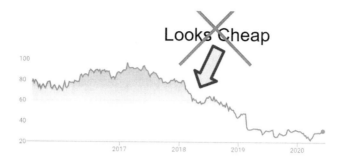

However, the stock was not actually cheap. Over the next few years, the stock fell pretty hard.

The Monte Carlo Fallacy

On August 18, 1913, there was a group of gamblers at the Monte Carlo Casino taking their chances on a roulette table. Some of the gamblers noticed that the ball landed on black 7 times in a row (the ball can land on either red or black). The ones who noticed that the ball had been landing on black said to themselves that "hey, it has been landing on black. It is going to land on red next." They betted a lot of money on red. It landed on black. Now, they thought that it had to land on red. They betted even more on red. It landed on black. Now, even more people were picking up on the pattern. Even more people were expecting it to land on red. "It must land on red because it landed on black for the past 9 times." Once again, they betted on red and lost, and more people picked up on the pattern and decided to bet on red again.

That night, the ball landed on black 26 times in a row. The gamblers collectively lost millions of dollars that night, but their story is going to save investors millions. Learn from their mistakes and don't let their loss go to waste.

The Monte Carlo Fallacy or the Gambler's Fallacy says that past events do not predict future events. Just because the ball landed on black 5,6,7, or 26 times does not make it any more likely that the ball will land on red the next turn. There is always a 50-50 chance on whether it will land on red or black. A coin, where both sides weigh equally, will always have a 50% chance of landing on heads and a 50% chance to land on tails, even if the coin has landed on heads for the past 20 coin flips.

The outcomes of past events do not affect the outcomes of future events. The same is true with stocks and technical analysis. Day traders (not investors) analyze stock charts and look for patterns. For example, the hammer candlestick pattern, in simple terms, is a pattern in the stock price's movement which, after a price declines, predicts that it will go back up; if a day trader sees the hammer pattern, the trader can expect the price to go back up.[60] However, this sort of technical analysis is also guilty of the Gambler's Fallacy. Just because a candlestick pattern was followed by an increase in stock price does not mean that the same pattern will lead to the same result in the future.

When people hear of people day trading, it tends to create a lot of excitement. It creates an atmosphere that sounds like it can be easy money if you can just follow a few simple rules and formulas. But, it is

---

[60] Day traders analyze stock charts. They look for patterns in the chart. Then, they use those patterns to trade stocks when they think that the same pattern is going to occur again. This sort of analysis is called *technical analysis*, and they usually analyze "candlestick charts" when looking at patterns. They look like a normal stock chart that shows the price of a stock, but it has a little more information in it. Day traders, in almost all cases, buy and sell their shares within the same day. They are not long-term investors. They look to make frequent, quick, small gains so that it adds up over time.

important to understand that the odds are not in your favor with day trading because historical patterns do not predict future stock prices.

The only factor that determines a stock's price is the supply and demand for that particular stock. As more people buy a stock, its price increases. As more people sell a stock, its price decreases. Candlestick patterns do not predict whether a majority of investors are going to decide to place an order to buy or sell a stock. Just because a pattern occurs, investors are not going to make a trade in order to make the predicted outcome of that pattern true.

It's important to be aware of the Monte Carlo fallacy when studying the stock market. There are many courses and books on the market where people pick out isolated patterns and use it to prove the effectiveness of a pattern. Sometimes they will pick out many instances of the same pattern in order to prove their point.[61]

Whenever you hear about someone using chart patterns to make trading decisions, I want you to think back to the gamblers in the Monte Carlo Casino on August 18, 1913. Please don't let the excitement of day trading overtake your logic and prevent you from making logical

---

[61] A candlestick pattern has realistically occurred many times due to the fact that there are so many stocks on the market, so many hours in a day when the markets are open, and so many days when trades are being executed. It is not hard to find that the same pattern has occurred multiple times. If there was 1 roulette table for every stock on the New York Stock Exchange, and if I threw the ball 390 times (one for each minute the stock market is open in a day) on each wheel and kept track where they landed, I would be able to find identical patterns taking place. For example, several wheels are likely to have a red, red, black, red pattern because there are so many opportunities for it to occur, like how the same pattern will occur many times in the stock market. But just because someone sees a red, red, black pattern, red is not anymore guaranteed to be the next color than if the pattern did not occur. It's always a 50-50 chance.

investing decisions. I have gotten emails from people who have lost hundreds of thousands of dollars day trading. I do not want the same outcome for you.

Although this short section on the Monte Carlo effect is not directly related to making investing decisions, almost everyone who gets intrigued by the stock market looks into day trading because its promises are irresistible. Let this section on the Monte Carlo Fallacy be a warning to you before you explore day trading (if you choose to explore it anyways because it's better to stick to what works and what has logical foundations).

Keeping Up with the Joneses

Arthur Momand created a comic strip called *Keeping Up with the* Joneses in 1913. In the comic strip, the McGinis family was working hard in order to climb the social ladder. However, their neighbor, the Joneses, was always ahead of them, and the comic strip's plot revolved around the McGinises trying to keep up their social status with the Joneses'.

The comic helped popularize a common phrase in 20th-century America, "keeping up with the Joneses." It represented the urge for people to buy something if their neighbors had it. If someone's neighbors had a new car, one would have the urge to buy a new car. Many people who tried to keep up with the Joneses during the 20th century ended up having a very difficult life. It led to many individuals living above their means and having too much debt. This phenomenon also led to many individuals feeling stressed out because they did not have what others had.

When you are investing, you don't want to be like the McGinis family. The McGinises always worked to buy what the Joneses had. If you see an investor who you admire, whether it be Warren Buffett or a YouTube guru, buying a stock, you should not automatically go invest in that same stock.

You have to keep in mind that large investors like Warren Buffett and mutual fund managers invest in very different circumstances than you do. Berkshire Hathaway (Warren Buffett's conglomerate) has a market capitalization of over $400 billion. Berkshire Hathaway and other mutual funds own so many companies because they have so much cash and need to diversify for their customers. So, any subsequent stocks which they are adding to their portfolio are unlikely to be as strong of a company as their initial stocks. In other words, they have already invested in stocks 1-10 which are on their list of wonderful businesses. Now, they are investing in less wonderful stocks, which probably would not even make it on their list of wonderful businesses. Now, they are not investing in the highest-quality businesses because they already own them. So, if you copy their most recent stock picks, you will not be investing in the highest-quality businesses.

Even if the Joneses are simply internet gurus or small investors, you should still not still "keep up" with them, even if they have a good track record because they are inherently biased. Bloggers and YouTubers make money by providing valuable and highly viewed content to their audience. These days, the most valuable content within the investment industry are videos and articles which recommend a person to buy a stock. These videos are popular because they help people make money. Who doesn't want to make money? In general, content creators rarely say that "this stock is not great" because that message is not valuable to their audience. People want to know where to deploy their money. They don't want to know where not to deploy it.

In order to make their content more appealing, many content creators tend to recommend more stocks than they should. For example, they may have a video or article called "the top 10 stocks to buy right now." In reality, only 1 or 2 of them might be a wonderful business at a wonderful price. Sometimes, they recommend stocks which I can easily say are not great investments. Usually, their recommendations don't outperform the market. Your goal should always be to outperform the market.

I'm not saying that there is anything wrong with consuming their content. They can open you up to new opportunities in the market which you were not aware of. You can find wonderful businesses at wonderful prices. You can learn about new trends in industries. You can also learn about other ways people like to invest in the stock market. However, you should not simply buy a stock because someone else has, like how the McGinises simply made purchases because the Joneses made a purchase. You **have to** conduct your own research before investing. Doing this will put you ahead of most investors. Don't let someone else's decisions define your finances.

How not to Sell Early

When you are investing in the stock market for the long term, it is inevitable that your stocks will go down. Even if you make the best investments in the world, on the way up, there will be times where your stocks will go down. You need to have the ability to not sell when they go down. *Even if you think that a stock is going to take a large hit soon, you should not sell it.*[62]

However, many people get scared and frustrated when their stocks' prices are expected to go down (i.e. someone might think that their stocks will go down because the markets are at all-time highs). So, they sell in order to maximize their **short-term** gains. Although selling before stocks go down can oftentimes maximize your gains in the short term, it is not a good idea for long term investing. In hindsight, those who sell out too early realize within a few months to a year that their stocks recovered fairly quickly after they fell and that the price has actually shot up higher than ever before. If you bought a wonderful business at a wonderful price, you would continue to make money by holding that stock for the long term, so you should not sell a wonderful business for short-term gains.

Similarly, I have noticed that some people sell many, if not all, of their shares when they see a large, quick gain. Once again, if the company still has a good, long-term outlook, they often realize that they missed out on larger long-term gains after they sold their shares.

A few years ago, I invested in Apple at about $153 per share. It went up pretty quickly, and I was up about 100% 1 year later. It was tempting to sell because I made a lot of money really quickly. However, I knew that I bought a wonderful business at a wonderful price. That wonderful business still had a strong, long-term outlook, so I did not sell it. Then it went down substantially. But I was patient, and it recovered. Now, at the time that I'm writing this book, I'm up about 150% on my Apple shares. The lesson here is to not sell your shares just because you are up a lot. A wonderful business bought at a wonderful price will

---

[62] The reasons to sell a stock will be outlined in a future chapter. If there is no reason to sell a stock you should not sell it, even if it is going to take a big hit.

continue to make even more money for many years. Don't sell too early!

I have noticed that this is mainly a problem with newer investors and that more experienced investors do feel the urge to sell but are able to hold themselves back. When you are investing, it is best that you never have an urge to sell early because that urge can create a lot of stress and make you sell too early.

In order to reduce the urge and anxiety, I have this very important tip: don't look at stock prices as often as you do. When you look at how your portfolio is doing on a daily basis, you will have one of two reactions. You will either become happy because you made money today, or you will become unhappy because you lost money today. The more money you make or lose, the more happy or sad you will become. When your stocks become volatile, you can have more extreme reactions. For many people, if they see a huge and quick upswing, they sell their shares and claim their gains. Similarly, if they witness or foresee a huge downward movement, they sell their shares, fearing that they will fall even more.

People sell their shares too early because they see their portfolio make large swings. So, the solution to not selling too early is to just not check your portfolio. If you do not see the large swings, you won't sell. Although it may sound absurd to not check your portfolio often, it is a very good strategy to prevent you from selling too early.[63]

I remember that, about a year ago, I was watching CNBC. They were interviewing an executive at Apple, who was also a very large shareholder. At the time, Apple stock was making a very large

---

[63] It is always good to keep up with major headlines related to your companies. I do not recommend that you stop consuming news in order to reduce anxiety. I suggest that you should mainly stop checking your stocks' current prices as often as you do.

downswing because many investors were worried about an American trade war with China and the fact that Apple was having trouble selling more units of higher-priced iPhones. The CNBC interviewer asked, "what are your thoughts on Apple's stock price?" The interviewee just said that he does not pay attention to Apple's stock price on a daily basis and instead focuses on the long-term potential of the company. He then said that he does not know what Apple's stock price is today.

    The stock was making headlines because of the dramatic moves it was making, and someone who has a large percentage of his wealth tied to Apple stock simply said that he did not care about what Apple stock did on a day-to-day basis. By not checking nor caring about Apple stock's short-term fluctuations, he had the capacity to hold Apple stock when it became volatile. Over the long term, he will be able to make tons of money in the stock.

    In addition to another strategy that will be explained in the next chapter about market capitalization, you can really help yourself by just not checking your portfolio often. By not monitoring your portfolio, you separate your emotions from the market. There is saying that goes something like this: looking at the plants in your garden does not help them grow, so looking at your portfolio does not make it grow.

Emotions Kill Your Returns: Contract

Your emotions are your greatest enemy. If you can't control them, they will control you. Your emotions will manipulate you when you focus on chart analysis, day-to-day stock prices, and other investors' stocks. So, in order to make good investing decisions, you must agree to the following:

I _____ *(insert name) agree to following about controlling my emotions:*

- I will not try to use past events in order to predict the price of a stock because that is not how the market works
- I will make my own investing decisions and not copy others. Content creators might be biased towards telling me to buy a stock, and large investors like Warren Buffett are likely not necessarily investing in the highest quality businesses today
- I will not constantly check my portfolio in order to see whether I am up or down because it will just create anxiety and stress which can cause me to make rash decisions

Market Capitalization and Opportunities: Summary

Market capitalization is an interesting topic. When I look at most investors' portfolios, I notice that they (intentionally or unintentionally) tend to invest heavily in only one market capitalization range. So, we are going to look at the pros and cons of larger and smaller cap stocks in order to determine where your money should be invested. We will also look at the role of a stock's market capitalization when investing and how you can avoid panic selling, which is common with small cap stocks.

Topics in this chapter include:

- Why there are more opportunities by investing in small cap stocks
- The reason many investors lose money when investing in small cap stocks
- How large cap stocks tend to be a very safe investment
- Why market capitalization does not really matter at the end of the day

Market Capitalization and Opportunities

"Buy a stock the way you would buy a house. Understand it and like it such that you'd be content to own it in the absence of a market." -Warren Buffett

As I learned how to invest in the stock market, I have found out that people have very interesting theories about which market capitalization ranges are the best to invest in. Some people decide to only invest in large and mega-cap companies, while others only invest in small and micro-cap companies. I have noticed that many people (including myself when I started investing) subconsciously invest in stocks that have a similar market capitalization, while other people specifically focus on diversifying their portfolio so that large and small-cap stocks represent a certain portion of their portfolio.

There are many contradicting strategies people have when it comes to market capitalization. Let's take a look at the role of a stock's market capitalization in a portfolio and the benefit of certain market capitalizations.

Classifications

For your reference, here are the market capitalization ranges for each of the classifications:

Classification	Market Capitalization Range
Micro-Cap	< $300 Million
Small-Cap	$300 Million - $2 Billion
Mid-Cap	$2 Billion - $10 Billion
Large-Cap	$10 Billion - $200 Billion
Mega-Cap	> $200 Billion

Why Invest in Large and Mega-Cap Stocks?

    Large and mega-cap stocks definitely do tend to have certain advantages over small, mid, and micro-cap stocks[64].

    One of the main advantages of these large companies is that you are likely to be a consumer of their products. You have probably purchased products from many large and mega-cap stocks. By being a customer of many of these companies, you know from experience which companies have a stronger brand, are better competitors, and have the strongest moat. You can use your life experiences to determine which businesses are the highest quality. Using your life experiences to determine which companies have strong moats is **one of**

---

[64] Since large and mega-cap stocks have very similar advantages, we are going to be referring to them collectively as large companies or large-cap stocks. Similarly, since the advantages of small, mid, and micro-cap stocks are pretty similar, we are going to be referring to them collectively as small-cap stocks.

the best methods to find high-quality companies. As a consumer, you understand which businesses you and other consumers favor.

The same is not true as often with small-cap stocks because most consumers and investors like you have never interacted with many small-cap stocks. You have never tried their products and services. The lack of being an actual consumer of these companies makes it difficult to assess their moat. It is hard to understand whether these companies are competitive in the marketplace unless you have actually done business with them.

I was recently looking at a small pet-food stock. It was very challenging for me to analyze the stock. As someone who was not a consumer of their products, I could not identify whether they had a strong moat. I knew that they had a unique brand and that they had intellectual property, but I could not determine how strong their moat was. As someone who has never purchased pet food, I did not know whether their brand and food were preferred over the competition, so I could not determine the strength of the brand. I ended up not investing in the company because I could not determine that their moat was strong.

Therefore, large and mega-cap stocks have an advantage because you are a consumer of these companies, allowing you to accurately analyze the quality of their moats.

Why Invest in Small, Mid and Micro-Cap Stocks?

Small, mid, and micro-cap stocks have advantages also. When compared to larger companies, small-cap stocks tend to trade at lower prices relative to their intrinsic value, especially during times of panic.

In general, large and mega-cap stocks are companies that have been operating for many years, have good reputations, and have a predictable future after times of fear and uncertainty (like a recession or global pandemic). If they are a dividend stock, many of the larger-cap stocks have been paying their dividends consecutively for decades.

Since larger-cap stocks are viewed as safer investments by many investors, they tend not to be volatile when there is fear in the market. For example, if there is a recession, every investor can virtually agree that almost all of the mega-cap stocks would make it through and persist without any long-term effects. These companies have been operating for decades, and many of them have dealt with recessions in the past, so most investors have confidence that they can persist through times of economic downturn. Also, most mega-cap stocks won't fail even if there was another Great Depression. In general, investors are less likely to sell their shares of large-cap stocks because there is just so much confidence in their long-term potential.

Many investors do not have the same confidence in smaller-cap stocks. Many of these stocks face lots of competition from larger companies, and they often do not have a predictable future because they are more prone to losing market share to industry leaders (if they have a strong moat though, they are unlikely to lose market share. This is why, as mentioned in previous chapters, moats are important). When investors start panic selling during a crisis, (which you should never do) do you think that they are going to sell their large-cap stocks (which they think are safe, stable investments) or their small-cap stocks (which investors think might be vulnerable)? In general, *since people are more likely to sell small-cap stocks over large-cap stocks, small-cap stocks can be found trading more at a discount than large-cap stocks. Finding them at a higher discount is the advantage of investing in small-cap stocks.*

Panic Selling

It is a good time to take a small break from talking about market capitalizations and instead talk about panic selling. Panic selling is simply selling stock when you see the price going down because you fear that it will go even lower. This is not a good thing to do because you will usually end up selling at a loss.

The main reason why people panic sell is because they don't know what they own. I was recently talking to one of my friends, and he asked me if a certain stock was good. I did not recognize the company.

"What does this company do?" I asked.

"It's a biotech stock," he said.

"What does biotechnology even mean?" I asked.

"It has to do something with living things and technology," he replied.

I could tell that he had no idea what the company did. He only knew the textbook definition of the industry that it operates in. It is such a huge mistake to not understand the company that you own. In fact, famous-investor Peter Lynch says that it is the most important thing:

> "The single most important thing to me in the stock market... is to know what you own. I am amazed how many

people own stocks [and] they would not be able to tell you why they own it."

Obviously, without understanding the business, you cannot determine whether it has a strong moat, whether it is better than competitors, and whether the industry is not rapidly changing. Just by investing in a company that is a better competitor, has a moat, and is in a slowly changing industry, you will be ahead of the vast majority of investors.

In addition to ensuring that you are invested in high-quality companies, knowing what you own prevents you from selling in a panic. When someone panic sells a stock, they sell it because they don't have confidence that the company will recover and persist over the long term. The only way to build confidence in the fact that the companies you own will recover and grow over the long term is by understanding and knowing your companies. If you don't truly know your stocks' products and services and why they are better than their competitors, how can you be confident that they will not be obliterated by the competition? Without believing in the long-term potential of your stocks' products and services, you will likely end up selling your stocks when they go down because other people are selling them. When you have confidence in your stocks and confidence in your calculation for intrinsic value, you could care less about whether they go up or down tomorrow because you know that you are going to make lots of money over the long term if you invested in it at or below its intrinsic value.

As I said earlier, small-cap stocks, in general, tend to have more opportunities than large-cap stocks because people tend to panic sell small-cap stocks more often than large-cap stocks. Why do you think small-cap stocks tend to be more oversold? Many investors don't even know and understand the actual business behind many small-cap stocks. Even if you go through the calculation for intrinsic value, you will lose confidence in the business during times of panic, unless you know the business that you are invested in and are confident about their competitive advantage.

Is a Certain Market Capitalization Better?

When I was talking about how large-cap stocks tend to be higher quality companies, I used words like "tend" and "in general" because these are only correlations. There are many exceptions. There are many large-cap companies that have weak moats, and there are many small-cap companies that have strong moats.

I also used words like "in general" when describing how small-cap stocks tend to become more undervalued than large-cap stocks. I have found small-cap stocks which were ridiculously overvalued, (even in a time of fear) and I have found large-cap stocks which were trading below their intrinsic value. So, just keep in mind that these are generalizations and that you will find many, many exceptions.

Before we conclude whether a certain market capitalization range is better overall, I want you to think about the following. Imagine you find two businesses and that both companies have a strong moat in their industry. Let's say that you are able to confidently predict the free cash flows of each of the stocks and you calculate each stock's intrinsic value. Let's say that stock A (a mega-cap stock) has an intrinsic value of $400 billion while stock B (a small-cap stock) has an intrinsic value of $500 million. If stock A and B are currently both trading at their intrinsic value, is one stock better than the other?

No! If two companies are equal in quality because of their competitiveness, and if both companies are trading at their intrinsic value, then they are both equally good investments. You'll roughly get the same return on investment from both companies, and neither of them is a risky investment due to their moats. Just because a large or mega-cap company is big, the money that it generates is not any more valuable. If you have two homes in the same neighborhood, one house that costs $100,000 and produces $5,000 in cash every year and another house that costs $200,000 and produces $10,000 in cash every year, neither of the homes is more undervalued than the other because you are getting the same return on investment.[65]

Therefore, market capitalization does not matter. You only need to follow the two key steps of investing: find a wonderful business and buy it at a wonderful price. The market capitalization of a stock does not make it any more wonderful, and a market capitalization is wonderful (no matter the size) if it can be justified by the future free cash flows that business will generate.

Since a company can be a wonderful investment whether it is a small or large-cap stock, it is important to analyze small **and** large companies. I know many people who only own large-cap stocks, and I know there are people who only look at micro-cap stocks. You want to make sure that you are analyzing large and small companies because wonderful opportunities can be found in both of them. If you don't, you are just going to end up missing out on a lot of investment opportunities. Who wants to miss a wonderful business at a wonderful price?

---

[65] Of course, keep in mind that this is a simplistic example. This example assumes that all other things are equal, ceteris paribus.

Market Capitalization and Opportunities: Contract

At the end of the day, if a stock is a wonderful business and is trading at a wonderful price, it is a good stock to buy, no matter the market capitalization. However, some people ignore analyzing certain market capitalizations. In other words, they ignore opportunities. So, I would encourage you to analyze companies of all sizes, as long as you understand the business.

I _____ (insert name) agree to following about market capitalization:

- I will always invest in businesses which I understand
- Investing opportunities can be found in all market capitalization ranges, and a stock's market capitalization is not always indicative of its potential risks or rewards
- A business is a wonderful investment if it is a wonderful business at a wonderful price, no matter the market capitalization
- In order to maximize the number of wonderful businesses that I can find, I will look for stocks with all kinds of market capitalizations

Take These Steps Before Investing: Summary

In addition to knowing how to invest your money, there are other steps that you need to take in order to become ready to invest. This chapter will go over how you need to prepare yourself on a personal financial level so that you can start getting rich through the stock market. This chapter is a little mundane, but it is equally important as other chapters in this book.

Topics in this chapter include:

- Key steps you need to take before investing
- Steps to reduce stress and anxiety when you are investing
- The ideal financial state to be in before starting to invest
- Why you need to prioritize paying off certain debts before investing
- The importance of an emergency fund

Take These Steps Before Investing

"Risk comes from not knowing what you're doing." - Warren Buffett

In order to become a successful investor, you need to be able to determine how much money you can invest in the stock market. You need to know an appropriate amount of money to invest given your financial condition.

Your Lifesaver

Before you invest in the stock market, you have to ask yourself the following question:

1. Do I have an emergency fund of 3-6 months' worth of expenses?

An emergency fund is essentially money that is just in savings. It is used only for emergency situations like losing your job. This is cash that should **never** be invested or used in non-emergency situations. It should only be used if you are in a situation where you have lost your main source of income. This is money that you can use to pay for your bills, rent, and groceries as you try to create or find a new source of income.

Personal finance experts like Dave Ramsey recommend that you save 3-6 months of expenses. For many people, it can take at least a few months to find a new job, especially when the economy is not doing well. This emergency fund will allow you to live if you are in a bad situation.

You never want to invest money that is a part of your emergency fund. The stock market can experience quick ups and downs, and it can be unpredictable at times. In fact, you can expect to be down on your investment in the short term because very few people end up investing at the very bottom; investing at the very bottom involves more luck than skill. If you don't have an emergency fund fully built, you are in a very bad position. If you lose your job and don't have an emergency fund, you could be in a position where you have to sell some of your stocks at a loss because you would need cash to pay for basic necessities. You don't want to set yourself up for stress and a potential loss on recent investment by not having an emergency fund built up. So, just make sure to have an emergency fund of 3-6 months of expenses.

Kill the Debt that is Killing You

Then, ask yourself the following question:

2. Have I paid off high-interest debt?

Whenever you decide to invest $1 in a stock, you are also making the decision to not use that same $1 to pay off debt. Sometimes it is just smarter to not invest in a stock and just work on paying off debt.

In an earlier chapter, we talked about how money is more valuable today because it compounds and grows over time. The same is true of debt. It grows over time. The more time you take to pay it off, the more money you will end up paying to your creditors. The more money you give to your creditors, the less money you will end up investing in stocks. So, it is important to pay off debts that grow quickly.

Before you invest in the stock market, you want to focus on paying off high-interest debts. These include payday loans and credit card debt. Although it is ideal to not have either of them, you want to pay those types of debts off before investing in the stock market. The annual interest rate on credit cards can be as high as 20%, and payday loans are even worse. An average stock market investor can get a return of 10%. By not paying off high-interest debt and investing in the stock market instead, you are going to end up losing money because the interest rate on those debts is very high; since the interest rate of those debts is higher than an average return on investment from stocks, paying off those types of debts would save you more money. Therefore, paying off those debts first will allow you to end up with more money than if you were just focusing on investing in the stock market.

You are probably wondering whether it is worth it to pay off low-interest debt before you invest. Low-interest debt is often paid over a long period of time. For example, car loans usually have a term of about 3 years, and mortgages can have terms as long as 30 years. If you solely focus on paying off these debts, you will miss out on years of compounding your money. Compounding is very powerful, especially in the later years of your life. Growing $10,000 with a 10% return would allow you to have $452,593.33 in 40 years. While compounding that same $10,000 with the same return over 41 years would allow you to have $497,852.65. One extra year of compounding would add an extra $45,259.32 in this scenario. The more money you invest **sooner**, the more money you'll make.

If you decide to not invest and just pay down low-interest debt, you are losing precious time. Low-interest debt like mortgages, student loans, and auto loans can take years to pay off, and time is money. Instead of focusing on solely paying off long-term debt, you have to balance paying off debt and investing so that you can take advantage of compounding over a greater span of time. Financial expert Dave Ramsey recommends individuals to invest 15% of their income for retirement purposes and then use the balance for paying off the mortgage and other debts. However, just keep in mind that the percent of your income that you need to save will depend on your current personal financial situation, and there is no one rule that will work for everyone. I'm good at finding wonderful businesses at wonderful prices, and I am not an expert in financial planning. So, if you need help determining how much money you need to invest for retirement, I would recommend talking to a financial advisor.

Being Smart with Your Money

Once you have paid off high-interest debts and created a good plan where you are able to contribute money to an investment account and pay off low-interest debts, ask yourself the following before you make an investment decision:

3. Do I need this money for any purchase within the next couple of years?

The words "purchase" can refer to anything. It can be a purchase as small as some groceries, or it can be a large purchase like the down payment on a new house. If you answered "yes" to this question, that money should never be invested in the stock market. If you need a certain sum of cash for an upcoming purchase, you should not invest it in the stock market because it may depreciate by the time that you need it. Especially during a recession, the stock market can be in negative territory for a prolonged period of time. You don't want money that you need for an upcoming purchase to be invested in the stock market. If you have money that you need soon invested in stocks, and if the stock market enters a bear market, your money could be less valuable at the time that you need to make the purchase. You would be forced to sell at a loss.

Instead of investing money that you need in the short term, focus on investing money that you really don't need for a very long period of time. When you are investing, you should always be intending to hold a company for the long term, so it makes sense to invest money that you would only need in a long period of time.

Being Happy While Investing

Once you ensure that you are investing for the long term, ask yourself the following question:

4. If I invest this money, will I still be living comfortably and happily in a situation where I lose most of it?

If the answer to the question is "no," then you should not invest that money in the stock market because you are emotionally attached to the money; you care too much about that money.

Once in a while, you will get unlucky and make a bad investing mistake. Warren Buffett invested in Delta Airlines however, fairly soon after his investment, all the airline stocks collapsed due to the COVID-19 outbreak. He, one of the best investors, ended up selling his shares at a loss.

It is important that you only invest money that you are not emotionally attached to. If you invest money that would cause you to be very sad if you lost it, you would end up very demoralized if you end up getting unlucky like Warren Buffett's investment in Delta Airlines. The stock market is a tool for building wealth and becoming a happier person. You don't want to let the stock market be a tool for anything except happiness.

On the other hand, if you only invest money that you are not emotionally attached to, you would not be consistently worried about whether a stock went up or down in the short term; an investor who is not emotionally attached to their invested capital does not care about the short-term performance of their investment. Not being attached to your invested money will help you focus on the long-term growth and potential of the company. Hence, it will prevent you from panic selling. Therefore, you should only invest money that you are not emotionally attached to.

If you think you are emotionally attached to all of your money, I would recommend investing 90% of your money in an index fund and 10% in stocks. This should allow you to have plenty of diversification so that your worst-case scenario is not too bad. If you want more information about index funds, read my chapter on diversification. But essentially, having most of your money in an index fund will allow you to never be too reliant on your stock picks.

This will help make you less emotionally attached to your money. For example, if you have $100,000 to invest, putting $90,000 in an index fund and $10,000 across 10 different stocks would mean that each of your stock picks would represent only 1% of your total portfolio. If, out of your 10 stocks, 1 of them was really bad and hit $0, your total portfolio would only be down 1%. At the end of the day, most people would not care if their total portfolio moved down by 1%. In fact, you can expect that to happen to your best investments during many days. So, by only investing a small amount of money into each stock and not relying on their performance, you would be less emotionally attached to each individual stock. Overall, this will help you focus on the long-term performance of your stocks and still be happy in a scenario where you are unlucky.

But of course, feel free to adjust the ratios as you wish. If you want, you could go much heavier in individual stocks as you become comfortable with investing. There is no set rule that works for everyone, so diversify and invest according to your own comfort level.

Take These Steps Before Investing: Contract

You never want your investments to hurt you because you were not in a position to start investing, so it is important that you agree to the following:

I _____ (insert name) agree to following about managing my personal finances:

- I should have an emergency fund of 3-6 months of expenses saved up so that I am never in a position where I have to sell my investments at a bad time
- It is best to pay off short-term, high-interest debt in order to create the shortest path to wealth. This should be done before I start investing or continue to invest
- I will not invest money that I need for short-term purchases.
- I cannot invest money that I am emotionally attached to. Doing this will prevent me from panic selling and will allow me to be a happier person

When to Sell a Stock: Summary

After learning when to buy a stock, the next logical question to ask is "when do I sell it?" So this chapter is going to give you 3 situations when you should sell a stock: you need cash for a better investment, the stock's cash-generating ability is lower than expected, or the company starts to lose its moat. This chapter will also go over the benefits of never selling a stock.

Topics in this chapter include:

- How never selling a stock is beneficial for your portfolio, in most cases
- Why Warren Buffett aims to never sell a stock
- The only 3 situations when you will need to sell a stock

When to Sell a Stock

"Our favorite holding period is forever." -Warren Buffett

When you are investing, you should buy a company as if you never intend to sell it. Buying a company at its intrinsic value will allow you to outperform the market over the long run. However, I'm not saying you should never sell a stock either. There are situations when you should sell a stock. First, let's take a look at 3 situations when you **must** sell a stock.

Situation #1: An Ever-Lasting Black Swan Flies in

When calculating the intrinsic value of a company, we were "forecasting future cash flows and discounting them at an appropriate rate," as Warren Buffett says. When calculating the intrinsic value of a stock, we are predicting that a company will produce roughly $X in free cash flow or dividends every year. However, a black swan can occur, and make your estimation inaccurate.

    In a book called *The Black Swan,* Nassim Taleb argues that there are unpredictable events, black swans, that will occur in the future and that no one can possibly predict. Black swans have 2 important characteristics for the purposes of this book: black swans have a huge, widespread impact, and they are unpredictable. For example, no one could have predicted black swans like World War 1, the rise of the internet, and the COVID-19 pandemic. No one knew that they were going to happen until they actually happened. Black swans also affect businesses and industries. No one could have predicted that Uber was going to come in and totally disrupt the taxi industry. No one could have predicted that Amazon would be able to successfully offer free, 2-day shipping and outcompete others. No one could have predicted that many small businesses would fail due to a widespread pandemic in 2020 until it actually happened.

If an adverse black swan affects one of your stocks, and if it can potentially have a **long-lasting impact**, you have to sell the stock. For example, let's say that you invested in a stock, assuming that it would have a perpetual growth rate of 2%. Then, ten years later, something happened in the industry due to a black swan, and revenues and free cash flows were cut by 50%. If you believe that the decline is **permanent**, then you should sell the stock. The business would now be providing significantly less cash than you thought it would provide. In other words, it would no longer be generating plenty of cash relative to the price you paid for the business. So, if a **permanent** and **disastrous** black swan hurts one of your stocks, you must sell it.

In our times, we have seen many smart investors sell their shares in companies **because their cash-generating ability declined below what was expected**. For example, the retail industry has been growing at a very steady pace for decades. However, more recently, certain retail businesses have seen their sales plummet due to online shopping. As a result, smart investors have sold their shares of retailers who faced disastrous consequences as a result of the internet age. This allowed them to save money before the situation got worse and to invest their money in better companies.

Imagine that you found a business that you expect to produce $100,000 per year for the rest of your life. Assuming that you have a required rate of return of 10%, you calculate that its intrinsic value is exactly $1,000,000. You patiently wait until you find the company trading at a market capitalization of $1,000,000. Once you find the company at that valuation, you decide to invest in it. After holding this company for many years, a problem appears. A black swan comes in, and now your company is only going to generate $60,000 per year instead of $100,000 per year in free cash flow. Since the company is generating significantly less cash than your expectation, you must sell it. It would make more sense to put your money into a company that is going to generate much more cash relative to the valuation of the business.

Here is the key takeaway from this section; if a black swan adversely affects one of your businesses and will cause it to generate significantly less cash than you expected it to generate **over the long term**, you must sell it and deploy your cash into a stronger company with better long term prospects.

Even Warren Buffett follows this key takeaway. After the COVID-19 pandemic hit, he decided to sell his shares in Delta Airlines. A black swan hit, and Buffett thought that the company's long-term, cash-generating ability changed for the worse. In fact, he said that he sold his shares because "the world has changed" for Delta. In other words, the company's future prospects when he sold his shares were not the same as when he bought them. The cash that he thought was going to be generated will actually never be produced. Essentially, he realized that his calculation for the intrinsic value of the stock was off. If you ever realize that your calculation for intrinsic value was dramatically off due to a black swan, you need to sell your shares.

Situation #2: You Need Cash

Your goal when investing is to invest your cash in the places that will predictably get the highest return.

When there is panic in the market, investors become filled with joy because there are many more stocks than usual which are trading at or below their intrinsic value. This is great for you. Along with other investors, you would be in a situation where many companies are trading below the ideal price to invest in them. You would be getting nice deals.

When you are in this situation, you would try to buy the highest-quality companies trading at the cheapest price relative to their intrinsic value. For example, if you see one stock trading at its intrinsic value and another stock that is trading 10% below its intrinsic value, you would ideally allocate more money to the stock that is trading 10% below its intrinsic value because it is cheaper than a stock trading at its intrinsic value.

Let's say that you decide to invest in that stock which is trading 10% below its intrinsic value. You would be pretty satisfied and happy that you deployed your capital. However, the stock market can be very volatile during times of panic. The panic can create even better opportunities in the market. A few days or weeks later, it is possible to find a totally different but high-quality stock trading 30% below its intrinsic value. This would be a very good stock to buy. However, you might not have cash right now in order to purchase the stock.

To raise money and buy the stock, you could sell your shares of a different stock. Specifically, you could sell your shares of a company that was trading at a less attractive valuation when you purchased their stock. You could sell the shares of a company that you invested in when it was trading 10% below its intrinsic value. This would allow you to raise some cash. Then, with the cash, you could invest in the company that is trading 30% below its intrinsic value.

Even if you just invested in a company, it makes sense to sell it if you find an even better deal in the market. Investing is all about allocating your money so that you can get the highest returns without too much risk. So, if you just invested in a stock that is trading at its intrinsic value, and if you need to raise money in order to invest in a more undervalued stock, it makes sense to sell the company that you just bought.

Situation #3: The Stock Starts to Lose its Moat

Moat's are very important to the success of a business. They protect a stock's ability to generate cash. They make wonderful businesses wonderful. In fact, I would be hesitant to invest in a company that does not have a moat. But moats do not always last (although we always aim to invest in timeless and slowly changing moats).

Over time, industries change. New competition comes in and disrupts industries. Even industries that are the slowest-changing industries can face massive disruption at times. If a disruption causes your stock to lose its moat, then you should sell it. You don't want to be invested in a moat-less business because they are vulnerable to competition, putting their cash-generating ability at risk. If the future cash-generating ability of a company is at risk, then the investment is risky because you can't be certain whether you would get a good return on investment.

For example, back during the 1970s, newspaper companies had phenomenal moats. They were the go-to source for news. If you wanted to know what was going on in the world, you read a newspaper. They were a source of entertainment, and newspaper companies had many paying customers. Certain newspaper companies were very successful because they had a strong brand moat. Customers would always buy their newspaper from the same company because customers had built trust and a relationship with certain companies; the strongest newspaper companies built brands, making it difficult for less well-known newspaper companies to get as many customers. Back during the 1970s, virtually no one expected these companies to lose their moat. In fact, Warren Buffett has said the following:

> "In 1970, Charlie and I were looking at the newspaper business. We felt [that] it was about as impregnable a franchise could be found. We still think it's quite a business, but we do not think the franchise in 2002 is the same as it was in 1970."

However, the industry completely transformed as consumers started to get their information online. Many people stopped purchasing newspapers. The industry-wide trend caused newspapers like the Chicago Tribune and The New York Times to lose their moat and competitive advantage. They were no longer the designated source of information. Also, their brand no longer brought in customers. As a result of losing their moat, newspaper companies lost huge chunks of revenue over time. You don't want what happened to the newspaper companies to happen to one of your stocks.

So, if you notice that the long-term moat of any of your stocks is falling apart, you must sell out of it. The longer you hold onto it, the worse it will get. Without a moat, the company will have a hard time growing their free cash flows and competing in the market. It will become hard for you to continue to get a good return on investment. Why would you want to stay invested in a company, worrying about whether it will continue to be competitive over the next few years, when you could just invest in a company that has a strong moat and that you don't need to worry about?

Should you Never Sell a Stock?

Unless one of the 3 reasons that we just went over does occur, you should ideally never sell your shares of stock. Let's say that you have been invested in a company for the past 5 years, and you are up about 100%. In other words, your investment has doubled over the past 5 years. You might be thinking of selling your shares in order to claim your gains because you are up a lot. However, if the company has a good outlook over the next few years, there is not really any good reason to sell your shares because you'll likely get a good return on your money again.

Also, when we calculated the intrinsic value of a stock, we forecasted the future free cash flows over the company's lifetime. Then we calculated the intrinsic value of the stock assuming that it would be operating for many years into the future. So, it only makes sense to hold stocks for many years into the future if you estimated that it would provide a good return for many years into the future. Especially since good investing opportunities take time to become available, it does not make sense to sell a stock and wait for an investment opportunity.

In fact, Warren Buffett agrees with the idea that you should never sell a wonderful business. At the 1998 shareholder meeting, he said the following:

> "Well, the best thing to do is buy a stock that you don't ever want to sell"

Essentially, he is saying that you are aiming to invest in businesses that you never intend to sell. Never selling is just an easy way to make money in the stock market, and there is just not a good reason to sell unless you are in one of the 3 situations where it makes sense to sell.

Even if you are up a lot on a stock, you should not sell. I know people who are up hundreds of percent on their stock picks because they bought a wonderful business at a wonderful price years ago. If they don't sell, they will still be making high returns as they did in the past (even though there will be hiccups along the road). If they sell now, they would need to look for a new place to deploy their money. That is just redundant. Why would you sell your shares and get cash when you are just going to reinvest your money?

Unless there is a better opportunity available, it just makes sense to have your money in wonderful businesses that you bought at wonderful prices because they will continue to make lots of money.

When to Sell a Stock: Contract

Selling at the right time is very important because it will either make or break your long-term returns. You always must aim to hold stocks for the long term. Hence, you should rarely be selling stocks. Instead, you want to consistently be building up your portfolio.

I _____ (insert name) agree to following about selling stocks:

- I should only sell if an ever-lasting black swan affects my businesses, if I need cash for a better investment, or if my company starts to lose its moat
- It is crucial to invest in businesses which you don't plan on selling
- I should not simply sell my shares because I can claim a profit. Holding for the long term will allow me to claim a larger profit.

Conclusion

"So the writer who breeds more words than he needs, is making a chore for the reader who reads." - Dr. Seuss

This is the end of the book. So far, you have learned pretty much everything that you need to know in order to get rich through the stock market. In reality, investing is not a hard topic to understand.

When you know the 2-step framework of investing, whether to buy a certain stock at a certain price is an obvious decision.

This book may seem a little shorter than other books about the stock market. In reality, after reading many books about investing, I have realized that they have lots of extraneous information that is not necessary in order to be a successful investor.

There is so much information and theories about investing that, if I compiled them into this book, you would have something as long as the entire Harry Potter series. Had I told you about many different investing strategies and theories, I would be doing you a disservice because you would not have a clear direction on how to invest your money and because there is no reason to know many investing strategies when you only need to know 1 amazing strategy. So, this book provided you with only 1 amazing stock market strategy that is simply the best investing strategy because it provides high, predictable, and low risk returns. It has been used by the best investors, and it will continue to be used by them. If you want to become one of them, it is time that you start implementing this 2-step strategy!

Acknowledgements

Special thanks to…

Ahil Lalani and Faizaan Majeed for assistance in the creation of the book cover,

Badruddin Jiwani and Bilquis Sana Khan for proofreading this book,

John Blix for teaching me about depreciation,

Josh Koo for being an awesome teacher,

Jonathan M. Sauerbrey for finding a mathematical error in this book,

and to all Glenbrook South High School teachers for making my high school experience awesome!

About Danial Jiwani

Danial Jiwani wrote this book while he was 17-years-old. His goal is to provide aspiring investors with a direction to find wonderful companies and to determine a wonderful price. If you want to contact him, feel free to email him at Jiwanid@gmail.com.